SHAPING THE FUTURE:

NEW EXPECTATIONS FOR UNDERGRADUATE EDUCATION IN SCIENCE, MATHEMATICS, ENGINEERING, AND TECHNOLOGY

A Report on its

REVIEW OF UNDERGRADUATE EDUCATION

by the

Advisory Committee

to the

National Science Foundation
Directorate for Education and Human Resources

Melvin D. George, *President Emeritus, St. Olaf College (MN)*
Chair, Committee on the Review

Sadie Bragg, *Borough of Manhattan Community College (NY)*

Alfredo G. de los Santos, Jr., *Maricopa Community Colleges (AZ)*

Denice D. Denton, *University of Wisconsin - Madison*

Peter Gerber, *MacArthur Foundation (IL)*

Mary M. Lindquist, *Columbus College (GA)*

James M. Rosser, *California State University - Los Angeles*

David A. Sanchez, *Texas A&M University*

Carolyn Meyers, *North Carolina A&T State University*
Consultant to the Committee

The EHR Advisory Committee

CHAIR: DR. JAMES M. ROSSER

VICE-CHAIR: DR. KERRY DAVIDSON

Dr. Susan Agruso
Director, Authentic Assessment
State Department of Education
Columbia, SC

Mr. Edward W. Bales
Director of Education, External Systems
Motorola Corporate Offices
Schaumberg, IL

Dr. Joan Barber
Director for Student Life
North Carolina School
of Science and Mathematics
Durham, NC

Dr. George Boggs
President
Palomar College
San Marcos, CA

Dr. Sadie Bragg
Acting Dean of Academic Affairs
City University of New York
Borough of Manhattan
Community College
New York, NY

Dr. Diane J. Briars
Mathematics Specialist
Office of Educational Design
and Assessment,
Pittsburgh Public Schools
Pittsburgh, PA

Dr. Kerry Davidson
Deputy Commissioner,
Academic Affairs and Research
Louisiana Board of Regents
Baton Rouge, LA

Dr. Alfredo G. de los Santos, Jr.
Vice Chancellor
for Educational Development,
Maricopa Community Colleges
Tempe, AZ

Dr. Denice D. Denton
Professor,
Electrical and Computer Engineering,
University of Wisconsin, Madison
Madison, WI

Dr. Charlotte K. Frank
Vice President,
Research and Development
The McGraw-Hill Companies
New York, NY

Dr. Alan J. Friedman
Director
New York Hall of Science
Flushing Meadows Corona Park, NY

Dr. Melvin D. George
President Emeritus
St. Olaf College
Northfield, MN

Mr. Peter Gerber
MacArthur Foundation
Chicago, IL

Dr. N. Gerry House
Superintendent
Memphis Public Schools
Memphis, TN

Dr. Jane Butler Kahle
Condit Professor
of Science Education
Miami University
Oxford, OH

Ms. Charlotte Keith
Indian Trail High School
Olathe, KS

Dr. Mary M. Lindquist
School of Education
Columbus College
Columbus, GA

Mr. Stanley S. Litow
Director, Corporate Support Programs
International Business Machines
Corporation
Armonk, NY

Dr. Jack R. Lohmann
Associate Dean,
College of Engineering
Georgia Institute of Technology
Atlanta, GA

Dr. Charles Merideth
President
New York City Technical College
Brooklyn, NY

Dr. Robert E. Parilla
President,
Montgomery College
Rockville, MD

Dr. Diana Garcia Prichard
Photoscience Research Division
Eastman Kodak Company
Rochester, NY

Dr. James M. Rosser
President
California State University
Los Angeles
Los Angeles, CA

Dr. David A. Sanchez
Department of Mathematics
Texas A&M University
College Station, TX

Ms. Maria Santos
Supervisor,
Math and Science Department
San Francisco Unified School District
San Francisco, CA

Dr. Robert Schwartz
The Pew Charitable Trusts
Philadelphia, PA

Dr. Calvert H. Smith
Office of Systemic Reform
State of Ohio
Cincinnati, OH

Dr. Gwendolyn W. Stephenson
Chancellor
St. Louis Community College System
St. Louis, MO

Dr. Uri Treisman
Department of Mathematics
University of Texas at Austin
Austin, TX

Dr. Leon Ukens
Department of Physics
Towson State University
Towson, MD

Ms. Donna L. York
Science Curriculum Coordinator
Anchorage School District
Anchorage, AK

St. Olaf College
Melvin D. George
President Emeritus

May 15, 1996

Dr. James M. Rosser, President
California State University - LA
5151 State University Drive
Los Angeles, CA 90032-8500

Dear Jim:

Enclosed is the Final Report of the Review of Undergraduate Education in Science, Mathematics, Engineering, and Technology, which I am now transmitting to you, as chair of the Advisory Committee to the Directorate for Education and Human Resources, National Science Foundation. This report is the product of more than a year of intense and comprehensive effort by the Advisory Committee's Subcommittee on the Review of Undergraduate Education, which it has been my pleasure to chair. The final report incorporates changes made by the Advisory Committee at its April meeting and is sent to you now, as authorized by the Advisory Committee, to conclude the subcommittee's work.

The review was conducted in accord with the direction given by Dr. Luther Williams (AD/EHR) in December 1994 and his subsequent charge to the subcommittee. Hundreds of persons and almost a hundred organizations have participated in and contributed thoughtfully to our review. We have drawn heavily on previous studies and analyzed large amounts of data relevant to our charge. We are deeply grateful to all who have helped us carry out this comprehensive assignment.

Special thanks are due to my colleagues on the subcommittee - Sadie Bragg; Alfredo de los Santos, Jr.; Denice D. Denton; Peter Gerber; Mary M. Lindquist; James M. Rosser; David A. Sanchez; and Carolyn Meyers (Consultant) - for their invaluable contributions not only to the shaping of the material into a report and recommendations but also for the rich ideas they brought personally to the review. The subcommittee's work was made easier by the effective assistance of EHR's Division of Undergraduate Education and its Director, Dr. Robert F. Watson, who responded promptly and thoroughly to all our requests for information and technical help. The subcommittee is grateful to the Advisory Committee for the valuable guidance given to us in discussions over the past year.

The subcommittee believes it has complied fully with the charge given to us, and we are unanimous in our strong support of the recommendations in the report. We hope that the title of the report, Shaping the Future, will indeed describe the impact of our work.

Sincerely yours,

Mel George

Melvin D. George
President Emeritus, St. Olaf College
Professor Emeritus of Mathematics,
University of Missouri-Columbia

cc: Dr. Luther Williams

CALIFORNIA STATE UNIVERSITY, LOS ANGELES

5151 STATE UNIVERSITY DRIVE, LOS ANGELES, CA 90032-8500

OFFICE OF THE PRESIDENT
(213) 343-3030

May 18, 1996

Dr. Luther S. Williams
Assistant Director for Education and Human Resources
National Science Foundation
4201 Wilson Boulevard
Arlinqton, VA 22230

Dear Luther:

On behalf of the Advisory Committee to the Directorate for Education and Human Resources (ACEHR), I am pleased to transmit the Final Report of the Review of Undergraduate Education in Science, Mathematics, Engineering and Technology. The report is the culmination of the review you asked this Subcommittee to carry out in your letter of December 22, 1994.

The review was an extraordinary effort conducted by ACEHR's Subcommittee on Undergraduate Education, under the leadership of Dr. Melvin D. George. It included very wide outreach; it engaged all parts of the National Science Foundation, at both early and advanced stages; and, it was monitored and guided through regular recourse to the forum provided by the whole Advisory Committee.

The final full and vigorous discussion of the text and recommendations of the report occupied much of the ACEHR meeting of April 3-4, 1996. The Advisory Committee voted unanimously to approve and endorse the report and to direct that it be transmitted to you. ACEHR requests that the report be presented to the Director and trusts that he will, in turn, transmit the document to the National Science Board.

This is a strong report, based on a thorough and perceptive review. The Advisory Committee is confident that it can be an effective blueprint for the next ten years of National Science Foundation leadership and programming at the undergraduate level. Further, its recommendations to other constituencies provide coherent guidance for continued improvement in service to important national goals in science, mathematics, engineering, and technology education.

Sincerely,

James M. Rosser
President

cc: Dr. Melvin D. George

NATIONAL SCIENCE FOUNDATION
4201 WILSON BOULEVARD
ARLINGTON, VIRGINIA 22230

OFFICE OF THE
ASSISTANT DIRECTOR
FOR EDUCATION AND
HUMAN RESOURCES

July 3, 1996

Dr. James M. Rosser
President
California State University - Los Angeles
5151 State University Drive
Los Angeles, CA 90032-8500

Dear Jim:

Thank you for sending me the Advisory Committee's report, *Shaping the Future: New Expectations for Undergraduate Education in Science, Mathematics, Engineering, and Technology*. I accept the report.

Your report is based on an unusually thorough review and is enhanced by its broad outreach to interested parties and sources of information and opinion. I appreciate the close monitoring of the review by the full Advisory Committee. Its detailed discussion of the findings and preliminary recommendations strengthened the report and provided additional confirmation as to the importance of its recommendations.

The report and its recommendations will undoubtedly receive careful attention by the many constituencies of – and stakeholders in – undergraduate science, mathematics, engineering and technology education. The recognition of both the Advisory Committee and the review panel that the review and recommendations could not be limited to NSF programs makes the report of much broader interest.

You expressed the confidence of the Advisory Committee that the report "can be an effective blueprint for the next ten years of National Science Foundation leadership and programming at the undergraduate level." Following careful examination and extended discussion within NSF, I will report to the Advisory Committee on how we expect to use that blueprint in NSF planning for the near and long term.

I appreciate very much the fine work of the committee on the review of undergraduate education. Please convey my thanks to Mel George, chair of the committee, for the impressive effort which he and its members carried out in response to my request of over a year ago.

Sincerely,

Luther

Luther S. Williams
Assistant Director

Telephone (703) 306-1600 FAX (703) 306-0399

Table of Contents

"NSF is determined that all students at all levels will be exposed to programs with high standards for understanding and accomplishment; that all students have the opportunity to advance to higher levels; that all students who enter advanced training at the professional level are well and broadly trained; and that the process of learning does not end with the classroom.

"Meeting this goal requires efforts from all parts of the Foundation. The undergraduate level plays a pivotal role."

— Excerpt from "NSF in a Changing World:
The National Science Foundation Strategic Plan"
(NSF 95-24, p. 29)

Shaping the Future: New Expectations for Undergraduate Education in Science, Mathematics, Engineering, and Technology

Executive Summary

Under the auspices of the Education and Human Resources (EHR) Directorate of the National Science Foundation (NSF), a committee of the Advisory Committee to EHR has conducted an intensive review of the state of undergraduate education in science, mathematics, engineering, and technology (SME&T) in America. The purpose of this review was to "consider the needs of *all* undergraduates attending *all* types of U.S. two- and four-year colleges and universities," addressing "issues of preparation of K-12 teachers in these fields, the needs of persons going into the technical work force, the preparation of majors in these areas, and the issue of science literacy for all." [Reference 1; list following Executive Summary]

This is the final report of the committee, which was to be "action oriented, recommending ways to improve undergraduate education in SME&T ... not just to the NSF but, as appropriate, to mission-oriented Federal agencies, business and industry, academic institutions and their faculties and administrations, professional societies, private sector organizations, state and local government, and to other stakeholders in undergraduate education." [1]

While the focus of this review was undergraduate SME&T education, that is just one part of the continuum of SME&T education in America that runs from pre-school through postgraduate work. The various parts of this continuum are interdependent; undergraduate SME&T education depends on the students who come from grades K-12, relies on faculty who come out of graduate programs, and prepares teachers for the K-12 system and students for graduate school. The kinds of programs offered for graduate students have significant implications for the future of undergraduate education; the professional standards adopted for student learning in grades K-12 impact undergraduate education as well. So, these sectors have mutual obligations to each other, and the fulfillment of those obligations is essential for the health of the whole.

Furthermore, as K-12 education changes, as a result not only of standards but of new emphases on inquiry, on active learning, and with new uses of technology, students will come to undergraduate education with new expectations, increasing pressure for reform at this level as well. And to sustain the kind of reform occurring in our nation's elementary and secondary schools, changes in undergraduate education, perhaps particularly in teacher preparation, will be essential. For all these reasons, this report and its recommendations are important to all parts of the continuum of SME&T education in the United States.

The goal – indeed, the imperative – deriving from our review is that:

All students have access to supportive, excellent undergraduate education in science, mathematics, engineering, and technology, and all students learn these subjects by direct experience with the methods and processes of inquiry.

America's undergraduates – ***all*** of them – must attain a higher level of competence in science, mathematics, engineering, and technology. America's institutions of higher education must expect ***all*** students to learn more SME&T, must no longer see study in these fields solely as narrow preparation for one specialized career, but must accept them as important to every student. America's SME&T faculty must actively engage those students preparing to become K-12 teachers; technicians; professional scientists, mathematicians, or engineers; business or public leaders; and other types of "knowledge workers" and knowledgeable citizens. It is important to assist them to learn not only science facts but, just as important, the methods and processes of research, what scientists and engineers do, how to make informed judgments about technical matters, and how to communicate and work in teams to solve complex problems. America's businesses and industry, governments, and foundations must provide active assistance and support in this critical endeavor. In an increasingly technical and competitive world with information as its common currency, a society without a properly educated citizenry will be at great risk and its people denied the opportunity for a fulfilling life.

The year-long review of undergraduate SME&T education leading to this report has revealed that significant change is occurring and that important and measurable improvements have been achieved in the past decade. Much of this progress is attributable to the leadership of the NSF, following the National Science Board's issuance of the 1986 report, "Undergraduate Science, Mathematics, and Engineering Education," NSB 86-100 ("the Neal Report") [2]. That report called for a significant program of support for undergraduate SME&T education and assigned primary, but not exclusive, responsibility for this activity to a separate division staffed by scientists, mathematicians, and engineers from many disciplines – now NSF's Division for Undergraduate Education – having the education of undergraduate students as its first priority.

The implementation of the 1986 Neal Report, despite funding of several key instruction-oriented programs at considerably reduced levels over what was recommended, has produced many positive results over the past decade. This success on the part of the NSF has reflected faithfully the dual mission of the Foundation in research *and* education and the conviction not only of the NSF but also of this committee that undergraduate SME&T education is the responsibility of scientists, mathematicians, engineers, and technologists alike.

Since the time of the Neal Report and the study that led up to it, the world has changed. The Cold War has ended, and public interest in and support of science have waned correspondingly. The use of new technology is exploding in all aspects of life. The economy is vastly different, with many fewer unskilled but high-paying jobs available to those without technical preparation. The demography of America and of its student population are changing dramatically.

Notwithstanding promising progress in SME&T education (many examples of which appear in the body of this report), much more remains to be done; there is now a broader and even more urgent agenda than there was in 1986. This message comes from the many contributors to this review: from focus groups of students and graduates, from testimony of employers, faculty, and administrators, from previous studies and surveys of all kinds. Despite the observation that America's basic research in science, mathematics, and engineering is world-class, its education is still not. America has produced a significant share of the world's great scientists while most of its population is virtually illiterate in science. Undergraduate SME&T education in America is typically still too much a filter that produces a few highly-qualified graduates while leaving most of its students "homeless in the universe." [3]

Too many students leave SME&T courses because they find them dull and unwelcoming. Too many new teachers enter school systems underprepared, without really understanding what science and mathematics are, and lacking the excitement of discovery and the confidence and ability to help children *engage* SME&T knowledge. Too many graduates go out into the workforce ill-prepared to solve real problems in a cooperative way, lacking the skills and motivation to continue learning.

Meanwhile, the world does not stand still. Knowledge keeps growing, new fields arise, other nations improve their educational systems, and new needs emerge. Governments at the state and federal level; business, industry, and the professional community; institutions of higher education; and the National Science Foundation, playing a key leadership role, must work together with a sense of urgency to make the necessary improvements. Students, for their part, must take *learning* more seriously.

The pressures on America's two- and four-year colleges and universities and on their students, facing an uncertain world of very constrained resources, are great. We do not ask for more of the same effort but rather for a more productive and rewarding kind of undergraduate SME&T education that produces long-lasting results more effectively and excitingly for both students and faculty.

The testimony of hundreds of participants in this review over the last year has led to a number of recommendations. These recommendations (detailed in somewhat different order in the body of the report) are for action to be taken by:

Institutions of higher education

We recommend that:

SME&T faculty: Believe and affirm that every student can learn, and model good practices that increase learning; start with the student's experience, but have high expectations within a supportive climate; and build inquiry, a sense of wonder and the excitement of discovery, plus communication and teamwork, critical thinking, and life-long learning skills into learning experiences.

SME&T departments: Set departmental goals and accept responsibility for undergraduate learning, with measurable expectations for all students; offer a curriculum engaging the broadest spectrum of students; use technology effectively to enhance learning; work collaboratively with departments of education, the K-12 sector and the business world to improve the preparation of K-12 teachers (and principals); and provide, for graduate students intending to become faculty members, opportunities for developing pedagogical skills.

Governing boards and administrators: Reexamine institutional missions in light of needs in undergraduate SME&T education; hold accountable and develop reward systems for departments and programs stressing the importance of SME&T education for all students; provide strong programs of faculty development; value and reward faculty who demonstrably facilitate student learning; reduce organizational rigidities, e.g., foster interdisciplinary work; make an institution-wide commitment to the preparation of K-12 teachers, in partnership with the schools; and support research and faculty dialogue on how undergraduates learn.

Accrediting agencies: Incorporate principles of sound undergraduate SME&T education into accreditation criteria, focusing on student learning, not just on organizational and process issues.

Business, industry, and the professional community

We recommend that:

Business and industry: Help those making public policy decisions understand the critical importance of quality SME&T education; make clear to educational institutions their expectations about graduates; and provide both partnerships and funding to colleges and universities to advance institution-wide reform initiatives.

National and regional media: Become better informed about the condition of undergraduate SME&T education in the United States and better inform the public about its critical significance for the nation's future.

Professional societies: Through journals and programs, honor and support education as well as research.

Publishers and testing agencies: Develop, validate, and disseminate materials and assessment tools incorporating desired goals for learning.

Governments at the state and federal level

We recommend that:

The President and Congress: Develop a new social contract with higher education, to put in place processes to sustain the relative excellence of the nation's higher education and so prepare the U.S. for a new century.

Other Federal funding agencies and foundations: Make strategic investments in support of the common agenda for improving undergraduate SME&T education.

State governments and higher education boards: Ensure that funding formulas and state policies provide incentives and rewards for excellent undergraduate SME&T education; and encourage collaborations among institutions, including sound articulation understandings.

The National Science Foundation

We recommend that the NSF:

- Make clear the high priority of undergraduate education. To do so, it is crucial to have within NSF a unit (the Division of Undergraduate Education, within EHR), staffed by practicing scientists, mathematicians, engineers, and technologists from many disciplines, that has undergraduate education as its *first* priority and that relates to *all* institutions providing undergraduate SME&T education. This Division must continue to maintain strong linkages with NSF's discipline-oriented research directorates, which in turn must continue to support undergraduate education within their specific fields.

- Aggressively improve undergraduate SME&T education through a variety of programs, beyond the base recommended in the Neal Report. A doubling of the present funding level in real dollars in the next decade will be needed to erase the gap between that recommended base and present program funding levels, and to extend the benefits of those programs to *all* SME&T students.

    1. First priority must be given to allocation of enhanced resources to the activities of the Division of Undergraduate Education (DUE) and to the undergraduate part of the *Alliances for Minority Participation* program in the Directorate for Education and Human Resources (EHR).

    2. The Foundation should encourage the research directorates to expand the allocation of their resources to discipline-oriented and interdisciplinary research-related educational activities that integrate education and research and that promote sharing the excitement of, and engagement in, research with undergraduates – with additional emphasis on primarily undergraduate institutions.

Further, we recommend that the Foundation:

- Lead the development of a common agenda for improving SME&T education.
- Give more priority to *implementation,* particularly of K-12 teacher preparation programs, faculty enhancement, and institutional reform, without diminishing support of innovative ideas and individual faculty curricular and pedagogical improvements.
- Lead the development of and provide support for a research agenda in human learning at the undergraduate level, using the results to evaluate programs (including long-term evaluation of student learning outcomes) and guide future program development.
- Develop an effective means of validating, codifying, and disseminating good practices in undergraduate SME&T education.

In all of its undergraduate programs, NSF should put emphasis on implementation of what is known to work, on genuine institutional change, and on sustainability of hard-won improvements. All of NSF's directorates should:

- Continue their support of strong activities to correct underrepresentation of women, minorities, and persons with disabilities among students and faculty at the undergraduate level;
- Support outreach activities that bring SME&T to the general public; and
- Consider funding mechanisms that both assign responsibility and provide incentives and rewards for achieving excellence in undergraduate programs not just to individuals, but to whole departments and entire institutions.

Our final recommendation is that the National Science Foundation accept leadership of the efforts necessary to implement all these recommendations.

This is an exciting agenda for action that concentrates on achievable goals. It requires a change in the way we think about SME&T education more than it calls for more hours or dollars spent on a task. It requires motivation as well as money, commitment as well as competence, and an interest in students as well as in science. Carrying out this agenda will be an energizing and exciting adventure in what is surely the most challenging and awesome enterprise in the world - human learning.

References and Notes for Executive Summary

1. Luther S. Williams, *Charge to the Subcommittee for Review of Undergraduate Education in Science, Mathematics, Engineering, and Technology*, Advisory Committee to the Directorate for Education and Human Resources, National Science Foundation (June 1995). [Copies available on request; See also Volume II of this report.]

2. National Science Board, Task Committee on Undergraduate Science and Engineering Education, Homer A. Neal (Chairman), *Undergraduate Science, Mathematics, and Engineering Education; Role for the National Science Foundation and Recommendations for Action by Other Sectors to Strengthen Collegiate Education and Pursue Excellence in the Next Generation of U.S. Leadership in Science and Technology,* (Washington DC: National Science Foundation, 1986, NSB 86-100).

3. Gerald Holton, Professor of Physics, Harvard University, *Introductory Comments*, National Research Council/ National Science Foundation Symposium on Science, Mathematics, Engineering, and Technology Education, Sponsored by the Exxon Education Foundation and hosted by GTE Corporation (Boston, MA: November 1995).

SHAPING THE FUTURE:
NEW EXPECTATIONS FOR UNDERGRADUATE EDUCATION IN SCIENCE, MATHEMATICS, ENGINEERING, AND TECHNOLOGY

I. To Begin: Background and Purpose of This Review

In 1986, the National Science Board (NSB) issued the report *Undergraduate Science, Mathematics and Engineering Education* ("the Neal Report")[Reference 1; list begins on page 73], which has guided the National Science Foundation's undergraduate education activities in the ten years since. Much has changed in the past ten years, and 1996 requires a new and even stronger vision of undergraduate science, mathematics, engineering, and technology (SME&T) education in this nation. This 1996 report, following an intensive review involving hundreds of thoughtful people across the country, conveys not only a vision but specific recommendations designed to bring it to fruition. This report is intended to be an agenda for action by the SME&T community and those who support it.

Improved SME&T education is central to shaping America's future.

It is the conviction of this review committee that improved SME&T education is central to shaping America's future. The future will increasingly require that citizens have a substantial understanding of the methods and content of science and technology – and some understanding of their potential and limitations as well as their interconnectedness. Furthermore, we believe that undergraduate SME&T education is the linchpin of the entire SME&T education enterprise – for it is at the undergraduate level that prospective K-12 teachers are educated, that most of the technical work force is prepared, and that future educators and professional practitioners in science, mathematics, and engineering learn their fields and, in many cases, prepare for more specialized work in graduate school.

On the basis of all that we have heard and learned during this review process, we urgently wish for, and urge decisive action to achieve, an America in which:

> **All students have access to supportive, excellent undergraduate education in science, mathematics, engineering, and technology, and all students learn these subjects by direct experience with the methods and processes of inquiry [2].**

This is a powerful vision of an America of the future where every person has an opportunity for a life of economic security and personal satisfaction through pervasive learning that provides competence in scientific and technical fields. This vision derives from the conviction that SME&T learning has value for its own sake as well as powerful utility in the workplace and in the exercise of citizenship.

We wish to help shape a future in which large numbers of students in America achieve substantially improved competence in science, mathematics, engineering, and technology fields, including better understanding of connections among disciplines and enhanced skills important for life as well as for work – problem-solving and lifelong learning skills, the ability to communicate effectively and work as part of a team, and personal traits such as adaptability, openness to new ideas, and empathy for the ideas of others. Our stress is on student learning that is measurable and involves much more than the acquisition of facts.

This vision focuses on *students* and on *learning*, but there are four other key words in the statement:

- **all** - every student should have access, whether in a two-year or four-year institution, not just those who intend to major in or pursue a career in SME&T; and groups traditionally underrepresented in SME&T (women, minorities, and persons with disabilities) must be included – for talent is not restricted to a pre-determined class of individuals;

- **supportive** – our programs must encourage and nurture students in subjects that for many seem forbidding and remote, if not impossible, and that have traditionally been viewed as the proper domain of only the few;

- **excellent** – we must have high expectations and provide superb educational experiences for every student, of sufficient quality that those who do major in these fields or otherwise go on to careers in scientific and technical fields are prepared at a world-class level; and

- **inquiry** – although there is disagreement about the meaning of the term "science literacy" and doubt about whether agreement is possible on a list of facts everyone should know, there is no disagreement that every student should be presented an opportunity to understand what science is, and is not, and to be involved in some way in scientific *inquiry*, not just a "hands-on" experience.

We heard elements of this vision from many people at many stages during this review. For example, in April 1995, approximately 300 faculty in science, mathematics, and engineering, together with administrators, representatives of professional societies, federal agencies, and foundations, gathered at the National Academy of Sciences (NAS) for a national convocation: *From Analysis to Action: Undergraduate Education in Science, Mathematics, Engineering, and Technology* [3]. Working under the auspices of the National Research Council (NRC) and the National Science Foundation (NSF), this group discussed for two full days the present state of undergraduate education and began the development of an agenda for significant improvements in the future. The one recommendation emerging from all the others and reflecting the conviction of this leadership group was almost identical with the vision above.

Such a vision is possible today only because of the enormous advances that have been made both in our understandings of human learning and in SME&T education in the past 10 years, many of them with the strong support of the National Science Foundation (see pages 17-27). In the area of learning, we know through research in cognitive psychology that the mind is active – it always interprets and is *not* simply a passive receiver of information "broadcast" to it. We know that students interpret new information in terms of what they already know; so, to promote learning, teachers must provide "stepping stones" for the minds of students to reach desired understanding [4].

In learning, the mind is active - it always interprets and is *not* simply a passive receiver of information "broadcast" to it.

We know that students rarely realize the applicability of knowledge from one context to another. We know that the diverse communities or cultures from which our students come have different values, norms, and expectations about the education process; learning is inhibited when those culturally-determined norms clash with what the instructor is doing. Research in sociology suggests that working in groups in a cooperative setting produces greater growth in achievement than straining for relative gains in a competitive environment.

Parallel with these increased understandings, the SME&T community has made enormous advances in undergraduate education in recent years, with the powerful support of the NSF – reflected particularly through its Division of Undergraduate Education, but in other divisions of the Directorate for Education and Human Resources (EHR) as well and with the effective participation of the research directorates. For example, the SME&T community has increasingly developed courses and curricula that stress inquiry, teaching effectiveness, and learning outcomes; has improved access to SME&T programs for those in groups that have traditionally been underrepresented in these fields; and has significantly increased the opportunity for undergraduates to engage in a real experience with inquiry/research.

Through the *Instrumentation and Laboratory Improvement* (ILI) program, faculty have been both stimulated and assisted in upgrading hundreds of laboratories in American colleges and universities, in connection with revamping courses to incorporate modern laboratory experiences. The NSF has helped institutions develop model teacher education programs, encouraged and supported collaboratives across institutional boundaries, and helped many undergraduate faculty enhance their competence. A major program, *Advanced Technological Education* (ATE), centered in the community colleges, has been initiated for preparation of the nation's technical work force. The level of conversation about pedagogy among faculty has increased, and many good practices and model programs have been disseminated; notable among these is the calculus reform effort, which is dramatically reshaping the way students learn calculus. All of these activities, stimulated largely by the recommendations of the Neal Report through programs designed and implemented by the NSF, have created a real momentum in SME&T education.

A research chemist from a major university recently testified about undergraduate education in her field: "The curriculum is knowledge for advanced studies. (I might argue it is knowledge for what used to be advanced studies). And yet 90% of these students will not be chemists. The classroom - it is embarrassing. Chalk and blackboard. There are hands-on experiments that the students can do. However, these are largely cookbook, and I think that although NSF really deserves a lot of credit for attempting to put instrumentation into these laboratories, I would say that still, at many, many institutions, my kitchen looks better than those laboratories. The textbooks . . . are large collections of facts. What I see really missing from these textbooks is the *process* of science. And finally, the exams . . . are really a nice way to give the student a grade, but I doubt that they really measure what the students are learning, where their critical thinking skills are."

But the data and the community – both those in SME&T fields and those outside who employ our graduates or influence public policy – say that there is yet a long way to go. The chancellor of a major research university, which is a member of the Association of American Universities (AAU) and a very large generator of scientific knowledge in many fields, recently said this: ". . . despite the outstanding character of American higher education, the one place where people see an Achilles' heel is the quality of science education."

Classroom learning must be accompanied by "knowledge derived from first-hand experience."

The 1993 report of The Wingspread Group on Higher Education, chaired by William E. Brock, was entitled *An American Imperative: Higher Expectations for Higher Education.* [5] While it criticized higher education generally, several of its points speak especially to SME&T undergraduate education. For example, it noted the 1993 National Adult Literacy Survey, which shows that a "surprisingly large" number of college graduates are unable to perform simple tasks involving mathematics. The report states that classroom learning must be accompanied by "knowledge derived from first-hand experience," a conviction that applies centrally to SME&T education.

Employers have consistently pointed out that higher education, because of the shortening half-life of knowledge, simply must do a better job of providing motivation and skills for life-long learning. And, as an executive of a large oil company testified at a recent NSF hearing: "Skills such as communications and teamwork are essential. Unfortunately, these are often given low priority during the SME&T professional's undergraduate education."

The president of a liberal arts college, at an NSF hearing recently: "A vital work force for the 21st century is peopled with the technically literate, inquisitive, and entrepreneurial in spirit. . . We have all talked about the need for improved educational experiences for our children. We have publicly acknowledged that our future leadership, tomorrow's work force, are today's children. Yet we do not adequately support the one profession in whose hands these children are. I am talking about teachers from K through college. NSF, with its dual mission of promoting the human resources as well as the discoveries, has a unique opportunity to make a difference."

A statement made by the National Science Board in 1994 [6] includes this sentence: "At the same time, the American public's level of scientific literacy and general technical preparedness are [*sic*] not adequate to meet the needs of the changing economy." That statement echoes the goal enunciated in the 1994 White House report *Science in the National Interest* [7] to "raise scientific and technological literacy of all Americans."

The president of an historically black institution spoke with passion at a hearing conducted as part of this review: "The intractable movement of African-Americans into the Ph.D. ranks, particularly in math, science, and engineering, is a moment of crisis for this nation. If every year we are having an erosion of those numbers, then you have got to ask what it is that we must do to get the feed system up to snuff so that more can come out of there. The inadequacy and obsolescence of laboratory facilities and the lack of modern technology on many campuses creates a drudgery syndrome with the teaching and doing of math, science, and engineering. So it is drudgery, and it is no wonder kids opt out of math and science and engineering."

And indeed they do opt out of SME&T. The extensive study *Talking About Leaving* [8] by Elaine Seymour and Nancy Hewitt notes the high attrition rates among SME&T majors -- for reasons having little to do with two popularly misconceived causes, namely language problems with foreign Teaching Assistants and large class sizes. Rather, the major reasons students identify for dropping out of SME&T have to do with the intimidating climate of the classroom, the poor quality of the educational experience (including too much dull lecturing and poor academic advising), the lack of encouragement for those interested in becoming K-12 teachers, the lack of motivation, inadequate counseling about career opportunities, and general lack of nurture of the student. SME&T education at the undergraduate level today is largely passive rather than active. It is certainly not providing "all students" access to "supportive, excellent" SME&T experiences that acquaint them with "the methods and processes of science."

A community college president: ". . . teachers too often discourage a student from pursuing a field or career, especially in math or science, by ignoring or redirecting them to "easier" studies. The future generation of scientists and technicians will be recruited by faculty who must democratize the process."

Thus, despite the enormous advances in undergraduate SME&T education in the past ten years, there is a challenge before us; it can be summed up in the words of David Goodstein: ". . . the United States has, simultaneously and paradoxically, both the best scientists and the most scientifically illiterate young people: America's educational system is designed to produce precisely that result. America leads the world in science - and yet 95 percent of the American public is scientifically illiterate." [9]

Both the preparation of teachers and the role of community colleges are much more central today among SME&T undergraduate education concerns.

Ten years ago the focus was on the problem of ensuring an adequate supply of world-class professional scientists for national needs. We must continue this important part of our responsibility for shaping the future. However, there is now a much broader agenda, with equally urgent new components, and it is in this light that the Directorate for Education and Human Resources of the NSF asked its Advisory Committee to undertake a new review of undergraduate SME&T education in the nation. First, the SME&T education community is coming to recognize what should have been clear all along - that the teachers of the students coming out of the K-12 system were prepared primarily *at the undergraduate level* for their school careers. Second, the national work force is changing dramatically, as high-paying but relatively unskilled factory jobs disappear in the face of foreign competition and technological advances; consequently the educational needs of the prospective work force are now vastly different. For these two reasons, both the preparation of teachers and the role of community colleges are much more central today among SME&T undergraduate education concerns. In addition, we have awakened to the long-term disastrous consequences of leaving major segments of our society substantially out of SME&T. So, while much has been accomplished, there are new and important agenda items to be addressed.

At a hearing conducted as part of this review [10], the superintendent of a major urban school system commented about new teachers coming out of undergraduate programs: "Many new teachers arrive at their first assignments lacking sophisticated skills in writing, speaking, and computing. All new teachers should be able to use technology and adapt to its roles and applications. SME&T content is also essential for all new teachers. In their content areas, SME&T teachers should know more about the subject materials than they are required to teach . . . (and they) should have the benefit of sufficient practicum/internship experience before they graduate."

The review process has been overseen by a committee of the Advisory Committee to the Directorate for Education and Human Resources, charged by NSF's Assistant Director for Education and Human Resources, Dr. Luther S. Williams [11], to:

> **" . . . consider the needs of all undergraduates attending all types of US two- and four-year colleges and universities that provide undergraduate education in science, mathematics, engineering and technology. In particular, the review should address issues of preparation of K-12 teachers in these fields, the needs of persons going into the technical work force, the preparation of majors in these areas, and the issue of science literacy for all."**

In June, 1995, the review process formally began, with the sending of a letter from Dr. Williams to some 200 leaders in the scientific and industrial community, including professional societies, and other federal agencies. More than 150 responses provided a major part of the information considered in this review, but they were supplemented in several very important ways. In particular, the review was conducted in cooperation with the National Research Council; the April 1995 NRC-NSF Convocation, followed by the NRC's "Year of National Dialogue" [12] about undergraduate SME&T education, provided much rich material. The Foundation convened focus groups of SME&T students, graduates, and parents in the fall of 1995 [13], and hearings were held at NSF during those months for disciplinary faculty, institutional leaders, and executives of employers of SME&T graduates [10]. The research directorates of the NSF have contributed to the review in several important ways. Finally, at many meetings of scientific societies and professional associations over 1994 and 1995, the issues of the review were discussed and valuable comments and reactions gathered. This report is the result of this extensive process of consultation and review.

II. A Look Back: Recent History of Education Reform

It may be instructive to spend a moment to look back at the recent history of science education reform, which has, as with so much human activity, tended to be cyclical. For example, there was a flurry of reform activity after the 1957 launch of the USSR's *Sputnik* that awakened America to the fact that it was behind in "the space race" and galvanized the nation to make changes. Numerous projects were undertaken, many with NSF support, to develop new curricula and instructional materials, primarily for K-12, that were inquiry-oriented and in which students were to be active learners, not passive subjects.

Despite the fact that a residue of these notable efforts remains in the commitment to "hands-on" science classrooms, these reform projects did not result in systemic change in either K-12 or undergraduate education. In part, this failure was due to the fact that the preparation of teachers did not change fundamentally, and many K-12 teachers were simply unprepared to deal with the very different new materials. In part, simple complacence set in after the U.S. became the first nation to land on the moon, which was taken as a clear (!) signal that the problem had been solved, and presumably once and for all. Then, in 1981, funding for much of the educational effort at NSF, particularly for undergraduate education, was reduced drastically and almost fatally.

Complacence set in after the U.S. became the first nation to land on the moon.

This cycle of intense activity, particularly under an outside threat, real or perceived, followed by complacence, erosion of public interest, and shifting of societal priorities, is not at all uncommon and will likely be repeated. But it is important to note that the rest of the world changes as well, so it is essential to consolidate our domestic gains, to institutionalize improvements, and to establish a new and higher base for the next cycle. What is important now, as we look at undergraduate education, is to put in place processes that will sustain the relative excellence of U.S. education in a competitive world.

The next cycle in undergraduate education may have begun with the issuance of the Neal Report by the National Science Board in 1986 [1].

History of Undergraduate Programs at NSF Since the Neal Report

The report, *Undergraduate Science, Mathematics and Engineering Education* (NSB 86-100), of the 1985-1986 Task Committee of the National Science Board has been the principal guide for the restoration and evolution of NSF's undergraduate education activities since its acceptance by the Board in March 1986. The numerous recommendations in the Board Report fell into two categories:

A. Leadership

The central leadership recommendation of the National Science Board Task Committee was that the Foundation

". . .develop quickly an appropriate administrative structure and mechanisms for the implementation of these. . . recommendations. The focal point should be the [Education Directorate]; it should foster collaboration among all parts of the Foundation to achieve excellence in science, mathematics, and engineering education."

The Foundation established such a unit later in 1986, and the present successor of its evolution is the Division of Undergraduate Education (DUE) in the Directorate for Education and Human Resources (EHR).

In 1986, The Board Task Committee urged NSF to: "take bold steps to establish itself in a position of leadership to advance the quality of undergraduate education."

The Board Task Committee made several additional major leadership recommendations to the Foundation. It urged the NSF to:

"(1) *take bold steps to establish itself in a position of leadership to advance and maintain the quality of undergraduate education in engineering, mathematics, and the sciences.*"

"(2) *stimulate the states and the components of the private sector to increase their investments in the improvement of undergraduate science, engineering, and mathematics education.*"

"(3) *implement new programs and expand existing ones for the ultimate benefit of students in all types of institutions.*"

"(4) *actuate cooperative projects among two-year and four-year colleges and universities to improve their educational efficiency and effectiveness.*"

"(5) *stimulate and support a variety of efforts to improve public understanding of science and technology.*"

"(6) *stimulate creative and productive activity in teaching and learning – and (7) research on them – just as it does in basic disciplinary research. New funding will be required, but intrinsic cost differences are such that this result can be obtained with a smaller investment than is presently being made in basic research.*"

"(8) *bring its programming in the undergraduate education area into balance with its activities in the precollege and graduate areas as quickly as possible;"* and

"(9) *expand its efforts to increase the participation of women, minorities, and the physically handicapped in professional science, mathematics, and engineering."*

Substantial progress has been made in all these areas under the leadership of EHR, particularly DUE; to establish and support this unit was a wise decision, as can be seen clearly in the many good things that have happened in SME&T undergraduate education since. NSF, with leadership from DUE and with major involvement of other EHR divisions and the research directorates, has made its mark through the programs described below.

B. Leveraged Program Support

Undergraduate programs at the NSF in 1986 were limited to the *College Science Instrumentation Program* (CSIP) in the Directorate for Science and Engineering Education (SEE), EHR's predecessor directorate, and two programs operating across the several research directorates: the *Research In Undergraduate Institutions* (RUI) program, supporting faculty research in predominantly undergraduate institutions, and the *Research Experiences for Undergraduates* (REU) program.

The Board Committee recommended specific changes in this pattern of support. It envisioned that by FY 1989, NSF's budget would provide funding for Laboratory Development, Instructional Instrumentation and Equipment, Faculty Professional Enhancement, Course and Curriculum Development, Comprehensive Improvement Projects, Undergraduate Research Participation, Minority Institutions, and Information for Long-Range Planning.

Laboratory Development (LD) and Instructional Instrumentation and Equipment (IIE)

Over the decade 1986-1995, the *Instrumentation and Laboratory Improvement* (ILI) program, which may be the most widely known and most highly regarded of NSF undergraduate programs, embodied the objectives of both LD and IIE, mostly in the support of compact projects. Large projects in LD became possible under a *Leadership in Laboratory Development* component of the *Instrumentation and Laboratory Improvement* (ILI-LLD) program initiated in FY 1994 to support the development of national models for undergraduate laboratory instruction that undertake fundamental reform and improvement.

The need for larger-scale programming was recognized by the NSB Task Committee.

While the need for such larger-scale programming was recognized by the NSB Task Committee, the experience of nearly a decade of less comprehensive projects had to be gained before either the Foundation or its constituent communities were ready to undertake it.

Faculty Professional Enhancement (FPE)

The report of the NSB committee suggested a combination of instruction-oriented and research-oriented activities, and both kinds have been supported. Because there were only research-related undergraduate programs at NSF when the committee's report was written, it is clear that the emphasis recommended was strongly on *direct* improvement of instruction.

- The pre-existing *Research in Undergraduate Institutions* (RUI) program has grown steadily: RUI awards to faculty in primarily undergraduate institutions are now about one percent of NSF's total research budget.

- The *Undergraduate Faculty Enhancement* (UFE) program in EHR/DUE was funded first in FY 1988. UFE enables faculty members in all kinds of institutions to adapt and introduce new content into courses and laboratories; to learn new experimental techniques and evaluate their suitability for instructional use; to investigate innovative teaching methods; to synthesize knowledge that cuts across disciplines; and to interact intensively with experts in the field and with colleagues who are active scientists and teachers.

Course and Curriculum Development (CCD)

A majority of DUE's *Course and Curriculum Development* (CCD) program funds support proposals for introductory-level courses, curricula, and laboratories that address two general priorities: the development of multi- and interdisciplinary courses that will contribute to the scientific, quantitative, and technological literacy of all students; and, the encouragement of faculty in SME&T to take leadership roles in developing education experiences that enhance the competence of prospective teachers and encourage students to pursue careers in teaching.

The Directorate for Computer and Information Science and Engineering (CISE), through its *CISE Educational Infrastructure Program*, supports innovative educational activities that transfer research results into undergraduate curricula in its fields. The projects are expected to act as national models of excellence. *Engineering Education Coalitions* (EEC), begun in FY 1988 as a joint effort of the Education and Engineering Directorates, became an Engineering Directorate responsibility in FY 1991. These coalitions stimulate bold, innovative, and comprehensive models for systemic reform of undergraduate engineering education, based on substantive resource linkages among engineering colleges and collaborating secondary schools.

- A joint effort of CCD and NSF's Division of Mathematical Sciences exemplifies the systemic effects possible when a clear and catalytic plan is funded over a period of years. Through a succession of solicitations (*Calculus Reform; Calculus and the Bridge to Calculus*; etc.), these NSF units supported a variety of experiments with new methods, pedagogies, and technologies for calculus instruction. By 1995, at least a dozen different innovative courses were being taught to at least 35 percent of all students enrolled in calculus.

A joint effort of the Course and Curriculum Development program and NSF's Division of Mathematical Sciences exemplifies the systemic effects possible when a clear and catalytic plan is funded over a period of years.

Comprehensive Improvement Projects (CIP)

The program of *Institution-Wide Reform of Undergraduate Education in SME&T* (the "IR initiative"), begun only in FY 1996, is the first clearly CIP-like undertaking of the Division of Undergraduate Education. As with the ILI-LLD program, considerable experience with smaller projects was essential before it was feasible to institute such a large systemic effort.

Undergraduate Research Participation (URP)

The NSB Task Committee's report recommended that FY 1986 expenditure on undergraduate research participation be increased several-fold. While the current support from NSF's research account for the *Research Experiences for Undergraduates* (REU) program has passed the financial target, important areas of the programmatic target have not been addressed. REU projects are concentrated heavily in research-rich environments, where possibly cutting-edge *research* may be, not where the *students* are. Such a pattern of awards leaves unfulfilled in most institutions a central purpose of URP as proposed by the NSB committee: ". . . the involvement of advanced undergraduate students in research in their colleges . . ."

Minority Institutions Program (MIP)

The Foundation has had programs designed to correct under-representation of minorities in science and engineering since the mid-1970's. Most were focused on individuals and were concentrated on the areas of graduate student support and faculty member research.

The focus of the NSB Report was on *institutions,* specifically minority institutions, but inferentially on non-minority institutions that were educating substantial numbers of minority undergraduates. EHR has two broad programs addressing these objectives: *Alliances for Minority Participation* (AMP, established in FY 1990), and *Research Careers for Minority Scholars* (RCMS, established in FY 1989 but consolidated with AMP in FY 1996). These programs fund efforts in student and academic enrichment, curriculum improvement, and institutional enhancement. An NSF-wide program, *Model Institutions for Excellence* (MIE), supports both instruction- and research-oriented activities.

Information for Long-Range Planning (ILRP)

NSF's *Division of Science Resources Studies* has principal responsibility to collect, study, and analyze information and data (e.g., on undergraduate science, engineering, and mathematics) to assist long-range Foundation planning. However, there is no general undergraduate SME&T education database at NSF - nor anywhere else, for that matter.

Other Programs

The Directorate for Education and Human Resources now has two substantial programs in areas not included in the recommendations of the Neal Report. They were developed in response to emerging needs, which have been reinforced by the findings of the current review:

- The *NSF Collaboratives for Excellence in Teacher Preparation* (CETP) program, sited in DUE, which supports efforts to achieve comprehensive changes in the undergraduate education of future teachers and to increase substantially the quality and number of teachers well-prepared in science and mathematics, especially members of underrepresented groups.

- The *Advanced Technological Education* (ATE) program, managed jointly by DUE and the Division of Elementary, Secondary, and Informal Education (ESIE), which promotes exemplary improvement in advanced technological education at the national and regional levels by supporting curriculum development and program improvement through specific activities of associate degree granting institutions and of alliances of two-year institutions with four-year colleges and universities, secondary schools, business, industry, and government.

Funding History of the Neal Report Recommendations

It is important to examine the funding history of these programs. [The approach taken in this section is to compare the level of funding recommended for FY 1989 in the Neal Report with the actual level of funding in FY 1995, both figures stated in 1995 dollars for comparability. In the case of one program recommended in the Neal Report but not started until FY 1996, we have reported the planned FY 1996 expenditure in order to give an accurate accounting of the current status of NSF undergraduate education programs recommended in the Neal Report.]

The NSF FY 1987 budget estimate (used in the Neal Report as a starting base for its recommendations) included $17.8M(illion) for undergraduate programming: $9.9M in SEE for the *College Science Instrumentation Program* (CSIP); and $7.9M in the research directorates for two efforts - $2.6M for support of the research of faculty in predominantly undergraduate institutions [mostly through the *Research In Undergraduate Institutions* (RUI) program], and $5.3M for the *Research Experiences for Undergraduates* (REU) program and similar activities.

- The NSB report recommended a total to be achieved by FY 1989 (i.e., in just two years) of $149.4M for undergraduate activities: $18.4M oriented toward research (12.3%), and $131.0-million oriented toward instruction (87.7%).

- NSF's budget in FY 1995 for the categories of program covered by the NSB report totaled $178.2M, of which $69.9M (39.2%) went for research-oriented activities and $108.3M (60.8%) for instruction-oriented programming.

While NSF expended $28.8M more ($178.2M less $149.4M) in FY 1995 than recommended by the Neal Report, this calculation masks serious deficiencies in several individual program areas. It is important, therefore, to examine the funding history in more detail.

- **Laboratory Development and Instructional Instrumentation** - after more than a decade, its funding ($21.7M in FY 1995) is less than a third of that envisioned by the NSB committee ($75.7M) - falling unit costs of computers and microprocessors have helped stabilize the prices of laboratory instrumentation, but demand for it has grown substantively. The intense competition for ILI awards (fewer than one-third of proposals receive awards) is but one demonstration of a great immanent need. *The funding gap in the LDI area is $54.0 million.*

- **Faculty Professional Enhancement (FPE)** – Current funding for instruction-oriented Undergraduate Faculty Enhancement is $7.2M, about 26 percent of that provided through the RUI program ($27.3M) to support the research of faculty in primarily undergraduate institutions, and far short of the $17.1M amount recommended by the NSB Committee for instruction-oriented FPE. *The funding gap for instruction-oriented FPE is near $9.9 million.*

- **Course and Curriculum Development** – Its current funding ($23.8M in FY 1995) in the Directorate for Education and Human Resources (EHR) is more than the NSB committee's target amount ($17.1M), and there is substantial support for such activities ($23.9M) in other directorates.

- **Comprehensive Improvement Projects** – FY 1996 first-year funding for the *Institution-wide Reform of Undergraduate Education* initiative ($4.0M) is a small fraction of that recommended for FY 1989 ($13.2M). *The funding gap in the Comprehensive area is at least $9.2 million.*

- **Research Participation** – The Neal Report recommended that the $5.3M budget estimate for undergraduate research participation in FY 1987 be increased by FY 1989 to $15.8M. NSF expended $27.8M for this purpose (*Research Experiences for Undergraduates* [REU] program) in FY 1995.

- **Minority Institutions** – The FY 1989 target was $6.6M; the FY 1995 expenditure on this category of programs was $42.5M.

- **Planning** – None of the FY 1989 target ($1.3M) has been funded.

It is clear that in three program areas (LD+IIE, instruction-oriented FPE, and CIP) current funding (FY 1996 data except for CIP) is some $73.1M below the level recommended in the Neal Report. This gap has limited seriously the ability of these programs to achieve all that was anticipated for them in 1986. However, current funding in four areas (CCD, research-oriented FPE, URP, and MIP) is some $103.2M above the level recommended in the Neal Report. This overage reflects differences in mechanisms and priorities from those embodied in the 1986 report.

These NSF programs, notwithstanding the gap in funding some of them, have produced clear benefits through institutions and faculty who have used NSF program support to improve the learning experiences of SME&T undergraduates. There has been a special emphasis on evaluation of the programs in undergraduate education at the NSF over the years. The process of evaluation is lengthy; results are not usually available until a program has been operating for some years; and evaluation foci sometimes change as the work proceeds. It is only in the last few years, for instance, that EHR's evaluations have emphasized student learning outcomes – clearly an important focus. Nevertheless, evaluations have demonstrated many program benefits and have led to modifications that increased program effectiveness.

Curricular and Pedagogical Improvements

Program evaluations have documented the fact that a growing group of faculty in many types of institutions, drawing frequently on this NSF program support, have developed and tested an impressive number of curricular and pedagogical improvements in undergraduate SME&T education in recent years. As an important part of our review, leaders in the SME&T community were asked to indicate the major improvements in undergraduate education in these fields that had occurred during the past decade.

A growing group of faculty have developed and tested an impressive number of improvements in undergraduate SME&T education in recent years.

We begin by summarizing the improvements cited most frequently in their thoughtful letters. These include:

- incorporating new knowledge into lower level courses more rapidly and more thoroughly;
- introducing SME&T concepts by examining current issues for which most students have some personal context that are illuminated by SME&T knowledge – particularly *new* knowledge;
- organizing courses, or often course *modules*, to address real-world problems;
- developing curricula that expose students to key inter-disciplinary connections, and multidisciplinary perspectives stressing concepts as much as facts;
- focusing on *processes* (how to pose researchable questions, how to acquire information to address those questions, assessing the quality of information) at least as much as on the transmission of facts;
- using the vast computational power of modern personal computers and mathematics to explore SME&T concepts and illustrate properties of matter in ways that entice students;
- ensuring that students have frequent access to *active* learning experiences, in class (such as in peer groups or in laboratory classes) and outside of class (as in study teams, using interactive class bulletin boards, and/or in faculty research projects);

- developing curricula that embody some or most of thc above features, and that take full advantage of modern technology, particularly personal computers, multimedia materials, digital libraries, hypertext documents, and access to vast networked resources, including databases and activities on other campuses;

- improving ancillary skills (communication skills, teamwork, respect for ideas of others, cognitive skills, etc.) as a critical byproduct of modern approaches to teaching and learning;

- ensuring that students have ready access to people who can provide them with reasonable assistance (faculty, teaching assistants, graduate students, advanced undergraduate students, and able peers);

- demonstrating respect for students' genuine efforts to learn, understanding that many learn through initial failures, and encouraging further efforts to learn;

- mentoring students, when this is possible; and

- devoting more energy to advising students about course selections and career options.

These improvements are attempting to nurture a sense of wonder among students about the natural world, while equipping them with tools to explore it and to learn.

A simple précis is that these improvements are attempting to nurture a sense of wonder among students about the natural world, to maintain students' active curiosity about this world while equipping them with tools to explore it and to learn.

Do these approaches really make a difference? Overwhelmingly, professional educators and researchers answered "yes." As one example, Alexander Astin, Director of the Higher Education Research Institute at UCLA, wrote to us about the findings of his comprehensive study of students and faculty, conducted several years ago [14]. He found extensive evidence documenting the importance of these kinds of curricula and pedagogy. Based on student responses and faculty interviews, Astin recommended that institutions should take several steps in order to make their SME&T programs more attractive and stimulating to a broad base of students. Institutional leadership appears to play a crucial role. In particular, his data suggest faculty will be much more likely to use active forms of teaching and learning if they work in an environment that encourages interdisciplinary work, team teaching, research opportunities for undergraduates, and high levels of faculty-student interaction; and that provides a supportive campus climate, with a high priority placed on undergraduate education.

The social sciences workshop we sponsored in February 1996 [4] described the research knowledge underlying the superior effectiveness of curricular and pedagogical approaches that are congruent with the types of improvements described above. One of the conclusions of the workshop was that a good teacher must do more than ensure the presentation of new and interesting information in the classroom – that teacher must also understand how the minds of students interpret and manipulate that information.

Stressing the importance of active student engagement, Professor Eugene Galanter (Director, Psychophysics Laboratory, Columbia University) observed in a letter to the review committee:

> **"Insofar as every science depends on data for both theory and application, laboratory or field data collection experience is an absolute necessity. Adding up numbers from a text-book example is not the same as recording those numbers or qualitative observations based on one's own effort. When students "own" their data, the experience, I suggest, becomes a personal event, rather than a contrived exercise. The tested retention of such information in our admittedly minimal current evaluations suggest enhancements in under-standing by a factor of six."**

Many contributors to this review noted that major efforts have been made to improve introductory courses and to develop valuable course sequences for students who are not planning to become research scientists or mathematicians, or practicing engineers. A number of these improvements require actions on the part of multiple parties – for example, the delivery of a multidisciplinary course for introductory students, or special courses built around major public issues by faculty from different departments. This shift to a broader student clientele is considered by these contributors to be particularly valuable for this nation as an increasing number of college graduates become "knowledge workers." As we have already noted, the SME&T education of students preparing to become teachers and technicians has received increased emphasis. A number of educators who wrote to us noted that there has been significant and sorely needed attention to these two groups of students.

Major efforts have been made to improve courses for students not planning to become research scientists or mathematicians, or practicing engineers.

Many contributors called our attention to the value added by partnerships among colleges and universities, as well as among faculty. Virtually all administrators and faculty from community colleges who wrote to us stressed the value of partnerships in improving the quality of education in community colleges. A number of our contributors, in discussing teacher preparation programs, noted the importance of partnerships between the education faculty and SME&T disciplinary departments, including participation by accrediting and licensing groups.

Virtually all administrators and faculty from community colleges who wrote to us stressed the value of partnerships in improving the quality of education in community colleges.

Professor David Hata of Portland Community College (Oregon) observed that the most significant improvement in undergraduate SME&T education that had occurred over the last ten years was: "Recognition of two-year colleges as *key players* in SME&T education as exhibited through federal legislation, increased funding opportunities through the National Science Foundation and other federal agencies, and inclusion in forums, conferences, and meetings to discuss undergraduate SME&T education. . . My ATE grant has promoted partnerships with colleges in New Mexico, Colorado, California, Arizona, and Oregon as we all endeavor to install associate degree programs in microelectronics technology, patterned after the Intel/Portland Community College program developed here in Oregon."

John Goodlad, who conducted a major study of education faculty in the US six years ago, and who is particularly interested in the education of K-12 teachers, wrote to us that the growing efforts to make SME&T fields more accessible to students who do not have a strong background in them, and are not planning to major in them, have collectively led to a major improvement among students planning teaching careers. Previously, Goodlad had written [15, p. 242]:

"Again and again, prospective secondary school teachers told me that they were unable to make connections between their undergraduate *subject matter* education and the high school curriculum they were required to teach. This is not a "methods" problem; it is a problem of understanding what curriculum reformers of the 1960's referred to as "the structure of the disciplines." The probability that few *teachers* graduate from college with the necessary understanding merely illustrates the probability that few college graduates of *any persuasion* do."

In his current letter, Goodlad wrote that he was particularly impressed with the success of efforts to promote accessibility through interdisciplinary learning over the last few years.

Goodlad stated: "This [success] has been [achieved] primarily through some relaxing in the boundaries separating these fields from one another. Through a greater focus on topics and problems cutting across the traditional subject fields, science and mathematics in particular have become somewhat more compelling for students."

The possibilities for such interdisciplinary learning are high in the immediate future, partly due to the fact that information technology increases the ease of such efforts.

A well-designed, active learning environment assists in the development of other skills (e.g. problem-solving, communications and teamwork).

Many curricular and pedagogical improvements are mutually reinforcing. Also very important is the observation made by many, particularly employers, that a well-designed, active learning environment assists in the development of other skills and traits they seek in employees: cognitive skills (problem-solving, decision-making, learning how to learn), social skills (communications and teamwork), and positive personal traits (adaptability and flexibility, openness to new ideas, empathy for ideas of others, innovative and entrepreneurial outlook, and a strong work ethic). This point has been made repeatedly in testimony at our hearings and in published studies and reports (see [16], for example).

The technology revolution has helped to accelerate much of the improvement in SME&T education. For instance, Doyle Daves, Jr., Dean of the School of Science, Rensselaer Polytechnic Institute, wrote to us:

> **"As the tools of the information, communication, and computing technology revolution become integrated into the educational process, the traditional classroom reliance on the lecture format becomes increasingly anachronistic. The essence of the new technology is the empowerment of the user. Inevitably, in use of computer-based technologies, learning becomes both active and under the control of the learner. Similarly, when students are actively engaged in the learning process using multimedia and information technology tools, almost inevitably they work together in groups or teams sharing insights and experiences and, in the process, learn teamwork, communication and organizational skills as well as subject matter."**

At the social sciences workshop (February 1996) mentioned earlier, Kenneth Foote, Professor of Geography and Associate Vice President for Research at the University of Texas – who both advocates and is developing a network of departments and faculty from many universities through the World Wide Web – noted that his students would frequently photocopy the electronic text that was part of their assigned homework until they realized that in paper form it was "dead," in other words, had lost the dynamic qualities provided by its hypertext links to other material. By the end of the course, students typically had become enthusiastic about the value of these links personally, very much in support of the Dean Doyle Daves' observation about empowerment.

Some Specific Improvements

There have been many specific curricular and pedagogical improvements over the last ten years. These are covered in greater detail in Volume II of this report. Here we will restrict mention to a few in order to provide a sense of what is being developed.

A large number of respondents identified calculus reform as perhaps the most mature curricular and pedagogical innovation over this recent past. Professor Hyman Bass of Columbia University (and Chair of the Mathematical Sciences Education Board) was one of many letter writers who nominated the calculus reform movement as one of the significant successes. This movement has achieved widespread adoption across every kind of campus, and has made calculus accessible to a much larger number of students. Overall, calculus learning is much less "from books," and much more akin to "an apprenticeship model."

Respondents identified calculus reform as perhaps the most mature curricular and pedagogical innovation.

Professor Bass observed that, "Originally, calculus reform was conceived as an effort to streamline and focus the content of calculus courses. It was expected that the emerging modules and textbooks, and the increasing pedagogical use of technology would be the most significant products." He went on to note that most people close to this activity now recognize that *new pedagogy* has been the most significant outcome. Teachers use these materials with considerable discretion and variation, not as tight scripts to follow. Of common concern to them are pedagogy and experimentation with its new forms: cooperative learning, open-ended problems, hands-on learning, and authentic assessment.

As often happens with educational improvements, there is currently a backlash to calculus reform [17]. Criticisms have been made that this new approach oversimplifies the subject, is too reliant on graphing calculators and personal computers, and has gone too far in making calculus look easy. While obviously there is need for long-term assessment of student achievement resulting from calculus reform approaches, even critics admit that the calculus reform movement has had a positive impact on the attitudes of mathematics faculty toward teaching.

An early and successful effort to improve physics instruction has been achieved by the *Workshop Physics* Project initiated at Dickinson College by Professor Priscilla Laws. *Workshop Physics* involves a redesign of the teaching methods in introductory physics courses to take advantage of recent findings in physics education research and to start students using modern computer tools. Integrated computer applications include microcomputer-based laboratory tools for data collection and display, spreadsheets for mathematical modeling and data analysis, and digital video analysis tools for the study of two-dimensional motion and electrostatics. Activity guides have been developed in order to support interactive teaching approaches in the traditional laboratory setting.

Workshop Physics is continuing to work with faculty at institutions with conventional course structures to incorporate interactive teaching approaches in their programs without restructuring the entire program. A number of institutions now use this approach, mentioned by several letter writers.

With NSF support, Louis Gross at the University of Tennessee (Knoxville) Department of Mathematics has developed entry-level and second-year curricula in biology emphasizing the great utility of quantitative approaches in analyzing biological problems, drawing on many examples from recent biological research. Software has been developed to allow students to experiment with a variety of biological assumptions by means of mathematical models and assess the resultant quantitative and qualitative behavior of these assumptions. This project was completed in the summer of 1995 and the software and approaches are still being developed and tested. The project has also been used as a springboard for workshops to bring together faculty working on similar efforts.

An example of efforts to bring new research results into the undergraduate curriculum is found in another curriculum project in biology. Jack Chirikjian of the Georgetown University Medical Center has been developing creative curriculum models to teach the core technology of biotechnology through theory and hands-on experiments applied in a context that demonstrates recent developments and applications of biotechnology in such areas as medicine, agriculture, forensics, and industrial procedures and processes.

NSF has provided significant support to the *BioQUEST* Curriculum Development Project through awards to Professor John Jungck at Beloit College. *BioQUEST* was started by a group of biologists who initiated the development of curricular and instructional materials around an approach to learning that stresses problem posing, problem solving, and persuasion of peers ("the three P's"). Initial work was followed by a demonstration project leading to the development of a network of biologists who are working to improve introductory courses and courses for students who are not majors in biology, by stressing open-ended problems in biology. The project has matured into major efforts to develop instructional materials, including additional software simulations and tools, materials that are not computer dependent, and materials that can be used in large lectures to engage students in problem-posing, problem-solving, and persuasion. The use of *BioQUEST* materials has been spreading.

The use of peer learning techniques in lecture courses is beginning to spread.

The use of peer learning techniques in lecture courses is beginning to spread as faculty become convinced of the effectiveness of this approach measured in student learning. One of the best known efforts of this kind is Professor Eric Mazur's *ConcepTests* and *Peer Instruction*, developed originally for an introductory physics course at Harvard University. In each course, the students' educational background and preparation are assessed, and each student's understanding of basic concepts is tested, in order to discover common student misconceptions. The results of this testing are incorporated into lectures that have three defining characteristics: brevity, emphasis on conceptual material, and active class involvement. In the active portion of class time, all students are required to try to explain concepts to one another in small groups. (Eventually, correct explanations are provided by the instructor and selected students.) Use of this technique has led to Mazur's conclusion (shared by others) that students are able to explain concepts to each other more efficiently than instructors – in all likelihood because they have only recently mastered them, are thus aware of the difficulties in understanding them, and know what to emphasize in their explanations.

More Things that Work

Another large scale effort supported extensively by NSF, headed by Art Ellis of the University of Wisconsin, focuses on extensive collaboration among chemists. It has led to the publication by the American Chemical Society of *A Materials Chemistry Companion to General Chemistry*. This "Companion" is comprised of text, problem sets, model kits, software, videotapes, and demonstration and laboratory experiments, showing how most topics covered in introductory chemistry courses can be illustrated with solids such as polymers, semiconductors, metals, superconductors, and ceramics - permitting reductions in the cost of laboratory courses. The project is stressing the adoption of these materials into introductory chemistry courses nationwide. Ellis has also worked in numerous other ways to enhance student learning of science and technology.

Multimedia approaches have been used more frequently in recent years. For example, Timothy Rowe, of the University of Texas' Department of Geological Sciences and Vertebrate Paleontology Laboratory, developed under NSF support multimedia software modules for a freshman level course for students not majoring in science that is called "The Age of Dinosaurs." This course now enrolls several thousand students annually at more than 20 colleges and universities in the U.S. The modules employ photographic-quality color images, 3-D models, animation, and sound as well as text coupled with hyperlinks that allow users to gain quick and easy access to other information that is remedial, basic, and supplementary - a clear advantage in classes where the incoming SME&T preparation of students differs widely. The modules are being published on CD-ROM for both Macintosh and IBM-compatible personal computers.

Another good example of multimedia-based curricula comes from an Advanced Technological Education project at CUNY Queensborough Community College. Professor Bernard Mohr in the Department of Electrical and Computer Engineering Technology is developing materials to support technology education in data acquisition, embedded systems, and multimedia, and high speed networks. These materials - which include networked laboratory manuals, text, and student exercises - are available also in a modular format to facilitate their use in technology education courses at other institutions. This project has placed a heavy emphasis on disseminating its results to other campuses through five-day workshops for faculty, and there is evidence that its use is beginning to spread.

Another innovative approach is introducing fundamental science in the context of major public problems, stressing their multidisciplinary nature.

Another innovative approach of high interest to students and relevant to their base of experience is introducing fundamental science in the context of major public problems, stressing their multidisciplinary nature. For example, NSF is supporting Zafra Lerman at Columbia College (Illinois) to work collaboratively with faculty at Indiana

University and Princeton University to develop a course built around environmental issues: "From Ozone to Oil Spills: Chemistry, the Environment, and You." Environmental issues offer an excellent vehicle to introduce the major scientific disciplines because many of them have important physical, geological, chemical and biological dimensions, while others feature aspects of engineering, economics, political science, and psychology.

There have been many NSF supported efforts in all of the major disciplines of science and engineering to improve the teaching of statistics – an important objective given the crucial role that statistical tools have in an information society. One of the early projects of this type was the development of a new introductory mathematical statistics course called *CHANCE*. Developed at Dartmouth College with the cooperative efforts of five other institutions – Grinnell College, King College, Middlebury College, Princeton University, and Spelman College – the course is designed to teach fundamental ideas of probability and statistics in the context of real world questions of current interest. Examples are statistical problems related to AIDS, the effects of lowering serum cholesterol on heart attacks, the use of DNA fingerprinting in courts of law, reliability of political polls, and the tendency of basketball players to shoot in streaks. Students learn how statistics are often manipulated, how to process information more effectively, how to ask the right questions, and are drawn into reading scholarly articles published in journals such as *Chance, Nature, Science*, and *Scientific American*.

The California Alliance for Minority Participation (CAMP) in Sciences, Engineering and Mathematics is one of the 13 sites supported by NSF under the AMP program, designed to reduce barriers to fuller minority student participation in undergraduate programs in the natural sciences and engineering. Organized around eight of the University of California campuses, CAMP is a faculty-based alliance operating statewide, involving the University of California, California State University, California community colleges and various independent colleges and universities, as well as corporate partners, national laboratories, and affiliated organizations. CAMP introduces students to SME&T fields through research opportunities and mentorships throughout their undergraduate education. The program consists of: (1) a CAMP undergraduate research scholars program, enabling minority students to participate in research with scientists at four-year colleges and universities in California, at national laboratories, and industrial research sites; (2) community college and pre-college alliances strengthening the preparation of minority students to pursue baccalaureate degrees in the sciences; (3) corporate alliances aimed at increasing the role of business and industry in the preparation of minority scientists and engineers; and (4) Alliance faculty symposia and colloquia addressing central issues in the sciences and in undergraduate education.

It is important to acknowledge the work of those disciplinary faculty who are working to expand our knowledge of how students learn.

Finally, it is important to acknowledge the work of those disciplinary faculty who are working to expand our knowledge of how students learn the subject matter of disciplines. The most promising and ambitious of these combine research programs on improved methods of promoting student learning with programs of instruction that impart this knowledge to future educators. A good example is the *Physics Education Group* at the University of Washington, under the leadership of Professor Lillian McDermott, which has received substantial NSF funding to develop their comprehensive, multi-faceted program as a model for other institutions to adopt.

And So?

The many strong curricular and pedagogical practices developed under the sponsorship of the Division of Undergraduate Education, and the wide variety and large number of institutions participating in DUE-funded programs, have improved the national prospects for comprehensive institutional reform, leading to revitalization of undergraduate education in SME&T disciplines and to greater attention and priority being accorded to undergraduate education. Recall that it was previously noted that the 1986 Neal Report [1] recommended that NSF should begin to encourage comprehensive reform – a recommendation that went unfunded and may have been ahead of its time. Recently, however, NSF's new IR program (*Institution-Wide Reform of Undergraduate Education in Science, Mathematics, Engineering, and Technology*), initiated in FY 1996 [18], drew more than 200 letters from two- and four-year college and university presidents indicating an intention to submit a proposal, and more than 130 formal proposals. This is a salutary result, considering that the aim of the IR program is to use its awards "to motivate changes in priorities and allocation of resources that will enable *institutions themselves* to support their reform initiatives." The fact that so many institutions felt ready to apply for an IR grant is further evidence of the advances made in undergraduate SME&T education in the past decade.

NSF's new *Institution-Wide Reform of Undergraduate Education* program drew more than 130 formal proposals - evidence of the advances made in undergraduate SME&T education.

The programs put in place by the NSF during these recent years have clearly provided important leverage and encouragement to take undergraduate education in SME&T fields more seriously and to make it more oriented toward active learning by students. The progress made by the SME&T community as a result, the broader interest in and commitment to change for the better in undergraduate education, and the enormous societal changes that have occurred in the last ten years have resulted in demands on undergraduate SME&T education almost unanticipated a decade ago and have led to the review we have undertaken.

III. The Situation Today: Findings of the Review

Education in America exists, of course, in the context of a particular society at a given moment in time. In the decade just past, there has been enormous change in the world external to education. These changes have profoundly influenced the climate for education and, in particular, this review of undergraduate SME&T education.

A Changing World and Economy

In this period, the Cold War has ended, and with it, there has certainly been some lessening of concern for science and technology, which have tended to have public support in times of an external national threat. At the same time, the use of technology has increased exponentially. Ten years ago there was virtually no Internet, no World Wide Web, and computers in classrooms were few and far between. Robots in factories were still something of a novelty. Today, however, there is an information technology revolution.

The economy has changed just as drastically. Manufacturing jobs have declined, while service and information-based segments of the economy have come to dominate. The economy is much more globalized, as corporations have become multinational in scope and global in their outlook. These corporations search the world for plant and office locations with cost-effective and productive workers. Their first concern is to increase their market value, and their major decisions are scrutinized frequently by investment analysts around the world. In many cases, high-paying jobs on American assembly lines for relatively unskilled workers have migrated overseas or are no longer necessary because of improved technology in the workplace.

Today there is an information technology revolution.

As a result of these changes in the economy, the work force in our nation is changing dramatically. More and more, the distribution of jobs is bifurcated, in the sense that jobs requiring relatively low levels of skill are paying low wages (primarily in the service sector) while those that provide decent economic opportunity demand skills far more sophisticated than those required by routine assembly line jobs in decades past. In the period following the late 1970's, annual income adjusted for inflation has fallen for typical workers who have not acquired at least several years of undergraduate education. The further below this standard of education workers are, the more their income has fallen. It is equally clear that those Americans who have technical skills will fare much better in the workplace of the world of tomorrow

Jobs that provide decent economic opportunity demand skills far more sophisticated than those required by routine assembly line jobs in decades past.

than will those who lack such skills and the educational preparation that helps produce them. Thus, SME&T education must play a new and expanded role in the preparation of the American work force for the next decade.

One of the realities of this changing economy is the fact that that lifelong learning skills have reached a stage of paramount importance. In an article in *The Wall Street Journal* entitled: "Consulting Giant's Hot Offer: Jobs, Jobs, Jobs," [19], Gene Wright of Andersen Consulting, seeking to underscore the need for experienced workers to take full responsibility to ensure their skills are up-to-date, stated: "There's a new skill set required. And it is the responsibility of the individual to retool, not the corporation." We must shape a future for America recognizing that the nation whose people are not well educated will lose out in the long run in this kind of economic world.

America's demography is also changing. SME&T and SME&T education have historically been the domain primarily of white males. Questions of value and equity aside – and they are *not* aside – the facts that the majority of Americans are women and that the proportion of Americans aged 18-22 who are members of racial or ethnic minority groups will rise in the aggregate from 25% in 1980 to more than 35% by the turn of the century (and that number is expected to rise above 40% by the year 2015) have profound implications for SME&T education. Unless SME&T education is much more inclusive than it has been in the past, we will be denying ourselves as a society the talents of the majority of our population. This is an intolerable situation – it is both morally wrong and economically foolish.

Unless SME&T education is far more inclusive than it has been in the past, we will be denying ourselves as a society the talents of the majority of our population.

There is no doubt that our society is now less committed to formal programs of affirmative action than was the case ten years ago. But the imperative described above – to be much more inclusive in SME&T education – is even stronger. And, while K-12 programming can expand the pool of those interested in pursuing careers in SME&T, it is at the undergraduate level where attrition and burnout can be most effectively prevented. What we in undergraduate SME&T education must do is to concern ourselves with *all* our students, not just those who historically have been represented in science, mathematics, engineering, and technology. Such a breadth of concern has important educational benefits as well, as it will force us to think more about how individuals learn and recognize what research has made clear: that there are differences in learning style which profoundly affect achievement. And let us not forget that increasing student achievement in SME&T education is exactly what is needed.

Rising Expenditures and Growing Financial Constraints

At the same time these general societal influences have changed so dramatically, there have been additional pressures on higher education. Public finances for higher education have become constrained for a variety of reasons (e.g. reduced growth in state tax revenues; competing demands from other areas, particularly K-12 education, the penal system, and delivery of health services), and there has been growing resistance to increases in tuition charges that exceed the rise in the Consumer Price Index.

Public finances for higher education have become constrained for a variety of reasons.

The cost per student of delivering undergraduate education rose from 1980 until the early 1990's, largely as a consequence of increases in the prices of educational inputs that have risen much faster than the CPI. According to national data, in the ten years ending in 1989, average education and general expenditures per student in all institutions of higher education grew at a rate 2.7% higher than the rate of inflation [20; p.300], and a careful estimate is that three-fourths of this above-average expense growth was due to rapid increases in the costs of instructional inputs, particularly faculty and staff salaries (which had declined considerably during the 1970's), rather than provision of more resources per student. Much of the remaining growth in expenditures financed sorely needed capital improvements and modern instructional equipment.

In public institutions of higher education, this extraordinary inflation had to be financed or accommodated by means other than raising public appropriations per student, due to the competing and pressing demands for public revenues from other sources. For example, consider the competition for public state-level revenues from the K-12 sector. Measured in constant 1994 dollars, expenditures per student in grades K-12 rose from about $4,000 in 1980 to nearly $5,500 in 1993, and state governments have provided a growing share of these public revenues over the last three decades - 40% in 1970, 47% in 1980, and nearly 50% since the mid-1980's. However, over this same period of time, public revenues for higher education, chiefly from state governments, remained in the vicinity of $4,500 per student in 1994 dollars, and have remained flat at 1.0% of Gross Domestic Product [21].

In both public and private institutions of higher education it was necessary to raise tuition charges faster than the CPI has increased in order to pay for the rapid inflation in the price of educational inputs. From 1980 to 1992, the sum of tuition for in-state students and "room and board charges" rose in constant 1994 dollars from $4,300 to $5,750 in public institutions, and from $9,920 to $15,700 in private institutions. Even allowing for the fact that 1980 was the low cost year, these annual charges are now substantially above their previous highs (reached in

1972) of $5,200 in public institutions and $10,850 in private institutions [21; year 1994, p. 182].

Average class size has crept upward at about 0.4% per year during 1979-89.

In most public institutions, and many private ones, the effects of these financial pressures have been noticeable to administrators and faculty: nationally, average class size has crept upward at about 0.4% per year during 1979-89 [20]; there is currently less variety in course and class offerings [22]; there is more incentive for (and greater resort to) hiring of teaching assistants and part-time adjunct faculty; and, budgets for student equipment, teaching laboratory facilities, and faculty development have been badly squeezed, according to our sources. Current financial constraints present major challenges and reduced opportunities in many institutions to try innovative approaches to undergraduate instruction while placing a premium on productivity-enhancing changes. These pressures were discussed by many who wrote to us.

Differences by Type of Institution

Current spending per undergraduate student now varies widely across different types of institutions.

These increases in expenditures per student after 1980 varied considerably among different types of institutions. Using the classification reported in *A Classification of Institutions of Higher Education* [23], both percentage and dollar increases were highest in highly selective bachelor's colleges and private research universities, and next highest in public research universities. In the remaining types of institutions, spending increases were either at the rate of increase in the CPI (in public two-year colleges) or moderately above this rate (in public and private doctoral institutions, master's institutions, and the remaining public and private bachelor's institutions).

One consequence of these uneven rates of growth in expenditures per student is that current spending per undergraduate student now varies widely across different types of institutions (although the data are not precise, particularly in large institutions with multiple missions). In recent years, the highly selective, private bachelors and research universities have been spending about 75% more per undergraduate student than public research universities. Public research universities have been spending about 20% to 40% more than other types of four-year institutions, and 60% more than public two-year colleges.

Changes in Higher Education

In addition to these financial pressures, there are other external influences complicating the picture for higher education. Governors, legislators, and parents in many states are raising questions about how undergraduates are treated, especially at major research universities with their heavy faculty research commitments and significant involvement of graduate assistants in the undergraduate instructional program. The whole debate over "teaching vs. research" in faculty workloads and in faculty rewards is being revisited by major controllers of educational purse strings.

Several universities are making significant changes in the internal culture surrounding undergraduate education, including in the way faculty are evaluated and rewarded. The American Association for Higher Education (AAHE) has taken a leadership role in finding and publicizing better ways to evaluate teaching effectiveness and in helping institutions think about revising promotion and tenure policies [24]; some of the disciplines (e.g., mathematics) have also given significant attention to such issues. But examples of such effort are few in number, are not yet pervasive in the higher education scene, and are counterbalanced by other forces: for example, some liberal arts colleges and comprehensive institutions are putting *increased* stress on research, on adding graduate programs, and on the acquisition of research grants by faculty. Not only do institutions need to make more changes in this important area, other stakeholders must assist them – for example, educational achievements would likely be accorded greater recognition in faculty reward systems if there were more effective ways of validating and disseminating such accomplishments, ways akin to the publication of research results.

The information technology revolution mentioned previously has been felt also by higher education. Faculty and administrators alike understand that more technology is needed on campus – to enhance productivity; to help to prepare students for the world they are entering (whether in the work force or as citizens); and to satisfy the demands of many entering students who are used to computers, the Web, CD-ROMs, and video, and who find learning from print and passive listening increasingly foreign.

Faculty and administrators alike understand that more technology is needed on campus.

Moreover, rapid changes in technology have created possibilities for organizations other than traditional colleges and universities to offer multimedia instructional packages of great sophistication and consumer appeal, independent of location. Some western state governors are proposing the creation of a "virtual university" using such technology, which might offer instruction more cheaply than a campus could. Whether or not such an institution can actually do all that a traditional campus can do in the human development process (which seems to require a social as well as an information context) is probably irrelevant; the fact that such technology offers the promise of cost reductions is enough to insure that such possibilities will be considered seriously by the political structures of this nation.

It is clear that large private corporations with major expertise in software and media could become serious "competitors" in the higher education arena in the near future. William Wulf, writing recently in *Issues in Science and Technology* [25] makes clear the importance of such an issue to higher education:

"Universities are in the information business, and technological developments are transforming that industry . . . Outside forces are always acting on universities. Some of them, notably the political ones, have great immediacy and hence get a good deal of attention. For example, university administrators are acutely aware of the . . . desire for greater "productivity" from the faculty, and so on. As important as these changes may be, I believe that information technology has a far greater potential to provoke fundamental change in our system of higher education. Moreover, I am certain that these changes are much closer than most people realize."

A paper by Eli Noam in *Science* [26; "Electronics and the Dim Future of the University"] observed: "Today's production and distribution of information are undermining the university structure, making it ready to collapse in slow motion once alternatives to its function become possible."

Noam notes that more articles on chemistry had been published in the previous two years than throughout all of history prior to 1900. With this kind of growth of knowledge has come increased specialization, which has led scholars in increasingly narrow fields to find electronic peers outside institutions that can no longer maintain all the subdisciplines. In addition, our students now come to the campus with electronic access to more information than is contained in the faculty, library, and laboratories of the college or university - and access to that information is available 24 hours a day at the student's convenience.

Noam continued:

"This scenario suggests a change of emphasis for universities. True teaching and learning are about more than information and its transmission . . . (are) based on mentoring, internalization, identification, role modeling, guidance, socialization, interaction, and group activity. In these processes, physical proximity plays an important role. Thus, the strength of the future physical university lies less in pure information and more in college as community; less in wholesale lecture, and more in individual tutorial; less in Cyber-U, and more in Goodbye-Mr.-Chips College. Technology would augment, not substitute."

Undergraduate Education Today

Turning now from external influences on higher education generally, we must examine the present picture of undergraduate SME&T education in America. Here, too, much has changed in ten years, even the answer to the question, "Where are the students?" For the students are, in fact, not in the kinds of institutions many people might guess [27, 28].

In the fall of 1992,

- **Two-year institutions** accounted for 44% of enrolled undergraduates, 41% of all undergraduate SME&T courses offered for credit, and 34% of all undergraduate SME&T course enrollments. Their share of SME&T enrollments is lower than either their share of SME&T courses or their share of total undergraduate enrollments, because nearly one-half of two-year college students are attending college part-time, and SME&T class sizes are relatively small - on par with those found in baccalaureate institutions.

In the fall of 1992, two-year institutions accounted for 44% of enrolled undergraduates, and 34% of all undergraduate SME&T course enrollments.

- **Research universities** enrolled 16% of all undergraduates in the fall of 1992, and had nearly 25% of undergraduate SME&T course enrollments. However, the number of SME&T courses taught for credit by research universities was only 15% of the total, because of a high frequency of very large classes.

- **Doctoral universities** accounted for 9% of all undergraduates and of SME&T courses, and for 10% of undergraduate SME&T course enrollments.

- **Master's institutions and engineering schools** with enrollments of 2,000 or higher accounted for 21% of all undergraduates, and for 22% of SME&T courses and SME&T enrollments; and,

- **Bachelor's institutions and small master's institutions** enrolled 10% of all undergraduates, 9% of those enrolled in SME&T courses, but 13% of all undergraduate SME&T courses offered for credit, reflecting a high frequency of small class sizes.

In FY 1992, 53% of undergraduates were enrolled in institutions that had *no* NSF funding for research or education, and another 12% were in institutions that received less than $100,000 from the Foundation. Only 25% were enrolled in institutions receiving more than $500,000 that year.

The picture is similar for all Federal agencies combined [29]. More than 88% of all Federal obligations to academic institutions for science and engineering in FY 1992 went to the 125 research universities. A second somewhat distinct group of 461 institutions received virtually all of the remainder (more than 11%) of this funding: this group comprised the 111 doctoral universities (with 6.2% of the total), the top 200 master's level institutions (with 4.2% of the total), the top 100 bachelor's colleges (with 0.7% of the total), and the top 50 two-year colleges (with 0.2%). Funding was also very highly concentrated within the doctoral and master's subgroups of this set of 461: the top one-eighth of each subgroup – 12 doctoral and 25 master's institutions – received more than 50% of the funds awarded in their subgroup.

"College" has become a nearly universal rite of passage in the 1990's.

And who the students are is changing, too, even though less rapidly than the demographic makeup of America is changing. "College" has become a nearly universal rite of passage in the 1990's. Overall, nearly 67% of female and nearly 60% of male high school graduates enter post secondary education within a few months of graduation. Another 10% to 15% of adults in their twenties will enter college a few years after graduating from high school, or after leaving high school and earning their general equivalency diploma [21, 30].

- At every age in the span from 18 to 24 years old, the percentage of US residents enrolled in colleges and universities has risen steadily since 1980. For example, 30% of 21-year olds were enrolled in college in 1980 compared to 40% in 1993.

- At older age levels in the interval from 25-34 years, there have not been increases in the *fraction* of the adult population enrolled in college during 1980-1993, but there have been growing *numbers* enrolled due to an expanding U.S. population in this age range. Overall about 8% of adults aged 25-34 were enrolled in college during this period – from 11-13% of 25-year olds and from 5-6% of 34-year olds. Equal percentages of the black, Hispanic, and non-Hispanic white populations in this age range were enrolled in college, but those enrolled were disproportionately women.

Due to these population demographics, undergraduate enrollments continued to rise through 1992, despite a drop in the number of high school graduates after 1988. [Undergraduate enrollments at the national level dropped in 1993, and began to drop slightly in large public institutions in 1992.] Most of the increase in undergraduate enrollments during 1985-1992 was fueled by rising numbers of part-time students (particularly women in two-year colleges) and older students.

The fact that undergraduate studies are being pursued by growing fractions of high school graduates and older "non-traditional" students means that the effort required of the SME&T faculty to reach all students with basic course learning experiences that will work for them has grown considerably in recent years.

From School to College

When these students arrive at America's two-year and four-year colleges and universities, they are (according to the higher education community) not well prepared for collegiate-level science, mathematics, engineering, and technology education. Analysis of high school transcripts does show some apparent improvement during the 80's in the SME&T courses taken by students during high school. The percentage of high school graduates taking the core curriculum recommended by the National Commission on Excellence in Education (4 units of English, 3 units of science, 3 units of social studies, and 3 units of mathematics) increased from 13% in 1982 to 47% in 1992 [22; *The Condition of Education*, 1994]. An increased number of students took more serious mathematics and science courses than had been the case. Specific gains made during 1982-92 were: 48% to 70% (geometry), 37% to 56% (algebra II), 32% to 56% (chemistry) and 14% to 25% (physics). But the levels reached in 1992 are still too low. For instance, only 22% had taken biology, chemistry, and physics, and only 21% had studied trigonometry.

High school transcripts show some apparent improvement during the 1980's in the SME&T courses taken.

In some states, high school graduation standards are being raised, and a number of major state universities have put pressure on the high schools by increasing course-specific entrance requirements. But, overall, the picture of the preparation of first-year incoming students for higher education is not a bright one – and the improvements that have occurred in the past decade have been overshadowed by the exponential growth of knowledge and the changing demands of society.

Many faculty in SME&T at the postsecondary level continue to blame the schools for sending underprepared students to them. But, increasingly, the higher education community has come to recognize the fact that teachers and principals in the K-12 system are all people who have been educated at the undergraduate level, mostly in situations in which SME&T programs have not taken seriously enough their vital part of the responsibility for the quality of America's teachers. The Neal Report [1] devoted one brief sentence to teacher preparation, for example (though much more to teacher enhancement). But, virtually every participant in the review work of this committee has expressed concern over the way the undergraduate SME&T education community is working in the preparation of teachers.

SME&T programs have not taken seriously enough their vital part of the responsibility for the quality of America's teachers.

With a more intensive and effective commitment on the part of institutions to the preparation of K-12 teachers, colleges and universities can raise their expectations about the preparedness of entering students. One way to do that might be for institutions to enter into "treaties" with the secondary schools providing that, after a certain date, credit will not be given at the collegiate level for remediation in SME&T.

Students in SME&T Classes

When the students arriving on our campuses begin to take SME&T courses, what do they find, and what do they experience?

As noted earlier, there have been a number of improvements on many campuses. There are more interdisciplinary opportunities; there are many more opportunities for some students to have research experience with a faculty member; and many more courses stress inquiry and feature active, collaborative learning. Students also find much more technology: on some campuses every student is expected to have a laptop computer; on many, each student has an e-mail account and can interact with faculty electronically. Unfortunately, none of these improvements is widespread and, overall, students report that their experiences in undergraduate SME&T education are not very positive.

Seymour and Hewitt's study produced findings that are very critical of faculty teaching practices.

Seymour and Hewitt's study, *Talking About Leaving* [8], produced findings that are very critical of faculty teaching practices.

Seymour and Hewitt conducted a large ethnographic study over the three-year period 1990-93 with 335 students majoring in the natural sciences and engineering (NS&E) drawn from seven campuses that were among the most productive contributors to the nation's flow of new baccalaureates in these fields. Most data were gathered by personal interview. Some data were obtained in focus groups of 3-5 students. An additional 125 students took part in focus group discussions on six other campuses. Half the students were in the biological sciences, physical sciences, and mathematics; the other half were in engineering. All of the students had SAT mathematics scores above 649 and, thus, were considered well-prepared to undertake NS&E studies in college. The student sample was designed to include slightly more students leaving (55%, all juniors or seniors) than remaining in NS&E majors (45%, all of whom were seniors). Underrepresented groups were over-sampled.

Generally poor teaching by the science and engineering faculty was by far the most common complaint of able students. Nine out of 10 one-time NS&E majors who switched to a non-NS&E major, and three out of four who persevered, described the quality of teaching as poor overall. The next most frequent complaint of NS&E majors was inadequate faculty advisement, mentioned by more than half of the successful majors.

Student condemnation of the faculty obsession with research changed dramatically when students were allowed to observe or participate in that research.

Students were very clear about what was wrong with the teaching they had experienced. They strongly believed that faculty do not like to teach (especially lower division courses); that they do not value teaching as a professional activity; and that they lack incentive to improve. In their explanations for the poor teaching they had experienced, students constantly referenced faculty preoccupation with research as the overt reason for their failure to pay serious attention to the teaching of undergraduates, and for specific inadequacies in attitude or technique. Student condemnation of the faculty obsession with research changed dramatically, however, when students were allowed to observe or participate in that research. The few students who had this experience liked the pleasant and open way in which faculty treated undergraduates in a research relationship, compared with their apparent indifference to them in a teaching context.

According to Seymour and Hewett, the perceived dislike of the natural science and engineering faculty for pedagogical contact with students cannot be simply explained by a greater interest in research, or by the bias of departmental reward systems. Students offered many examples of non-NS&E faculty who evidently enjoyed teaching, saw it as an integral part of their work, and took the trouble to do it well. Important elements in what students saw as good teaching were openness, respect for students, the encouragement of discussion, and the sense of discovering things together. Student comparisons of NS&E teaching styles with those in other disciplines are permeated with strong contrasts: coldness versus warmth; elitism versus democracy; aloofness versus openness; and rejection versus support.

The distancing of faculty from students was sometimes increased by sarcasm, degradation, or ridicule. These practices, apparently rare in non-SME&T courses, had the effect of discouraging voluntary student participation in classroom discussions, and created an atmosphere of intimidation.

Student criticisms focused on:

- Lack of student-teacher dialogue, which was thought also to reflect faculty indifference. Classes were mainly one-way lectures, which students compared unfavorably to the high school experiences of many of them, in which there was considerable dialogue.

- Evident poor preparation for lectures, indicating to students that faculty were disinterested in student learning. Students were particularly frustrated by faculty who seemed unable to explain their ideas sequentially or coherently.

- Students also wanted but typically did not find many illustrations, applications, and/or discussions of implications. Nevertheless, students did not believe there was anything intrinsically dull about NS&E class material, even though student interest in many classes began to flag when faculty failed to present material in a stimulating way. Many students made reference to the "monotone" voices and dry recitations of their instructors lecturing.

- Class tedium grew in instances where faculty were "over-focused" on getting students to memorize material.

- Students identified as *worst* practice reading or copying material straight from text books. Reports of this practice were common in every SME&T discipline and on every campus. Another version of this teaching style was sometimes referred to as silent teaching – an instructor writing on the board with his/her back to the class, whom he/she addresses minimally and infrequently.

- Seniors who were going to graduate in SME&T made it clear that the focus on weed-out objectives and use of poor teaching practices in the first two years had given them a shaky foundation for higher level work. They expressed resentment that their own education had suffered in the effort to discard others.

- Non-majors also expressed the opinion that their needs for basic understanding of science and mathematics had not been met in lower division SME&T courses.

The focus groups of students convened for the current review covered [13] a broader cross section of undergraduates than those created for *Talking About Leaving*; but, it is noteworthy that the opinions expressed about introductory courses were similar.

In our focus groups, students identified introductory SME&T courses as a major barrier.

In our focus groups, students identified introductory SME&T courses as a major barrier. Many non-SME&T majors were discouraged (or screened out) from pursuing further studies. SME&T majors found the introductory courses very challenging, and often described them as "weed-out" courses. All types of students objected to the large lecture format often used in these courses. (Students from two-year colleges, historically black colleges, and comprehensive institutions were not as negative about these courses as those from research and doctoral universities, which have the largest classes.) Even the recent graduates had no difficulty recalling the generally unpleasant experiences they had had in introductory courses.

- Students singled-out the practice in some large lecture classes of using television monitors in separate rooms to serve students who could not fit into the lecture hall as very discouraging. The perceptions of many students was that the faculty did not want to teach these courses.

- A significant number of students objected to the competitive atmosphere in introductory SME&T courses, calling it a barrier to learning.

- Seymour and Hewitt found the most frequent student complaint to be the often weak relationship between classes and supporting laboratory work. Students in our focus groups found similar fault; some found laboratory exercises to be mechanical – seemingly unconnected to concepts of science. Lack of faculty or teaching assistant expertise on site in the labs was cited as another weakness.

Students not only find problems in individual courses – they also experience a weaker curriculum. For instance, they find fewer laboratory opportunities because some institutions have made decisions to lower overall costs by reducing the number of laboratory sections and adding or substituting non-laboratory courses.

- Comparative data from two national samples provide us with the 12-year, undergraduate course-taking histories of two high school classes: the Class of 1972 (sampled by the National Longitudinal Survey, NLS) and the Class of 1982 (sampled by the High School & Beyond Survey, HS&B). These data indicate that laboratory courses offered and taken dropped by more than 20% from the 1970's to the 1980's.

- The 1993 National Survey of Postsecondary Faculty [28] indicated that only a small fraction of SME&T courses offered in the fall of 1992 were primarily laboratory courses. At the freshman and sophomore level, these fractions ranged from around 20% in the biological sciences, physical sciences, and engineering, to under 5% in the mathematical sciences and social and behavioral sciences. Only about 10% of students in the physical and biological sciences enrolled in the 20% of courses that were laboratory-centered, while 20% of engineering students enrolled in the 20% of courses so centered.

Only a small fraction of SME&T courses offered in the fall of 1992 were primarily laboratory courses.

Evidence of erosion in undergraduate science requirements was provided recently in a study conducted by the National Association of Scholars, *The Dissolution of General Education: 1914-1993* [31], in which the Association reported that 90% of 50 highly selective institutions required their students to take courses in the physical and biological sciences in 1964, whereas in 1993, only 34% of them maintained this requirement.

In addition to the data about current deficiencies in undergraduate SME&T education discussed above, we have reviewed the opinions of the many leaders in the SME&T community who provided input for our work. In the letter initiating this review, Dr. Luther S. Williams (NSF's Assistant Director for Education and Human Resources) asked not only that respondents identify improvements attained in the past decade but also that they comment on remaining barriers to further improvement, reflecting the needs of society [32]. Improvements and barriers were discussed also in the hearings at NSF in the fall of 1995 and in the review-focused sessions at many meetings of scientific and professional organizations in the past two years.

Barriers to Improvement

Some weeks before this review was initiated by Dr. Williams, he expressed concern about barriers to improvement of undergraduate SME&T education in the keynote address to an NSF-sponsored conference [33]:

"The [various types of] two-year and four-year . . . institutions have not yet responded substantially to the recognized need for cooperation and collaboration. Walls still exist between disciplines and academic units. These walls are ill suited to educating the many different individuals seeking preparation for a vast array of personal and professional goals in an increasingly complex world. These institutions have failed to prepare adequately for the new ways of learning that begin at the precollege level and must be continued at the undergraduate level. There is growing concern that what is taught does not adequately prepare students for the world they enter upon graduation. [Institutions] have not yet [developed] the potential of educational technologies fully, nor [applied] what is known from research on teaching and learning fully. We can no longer alter students to fit the abilities of educational institutions; we must alter the institutions to fit the needs of students."

"Walls still exist between disciplines and academic units. These walls are ill-suited to educating the many different individuals seeking preparation for a vast array of personal and professional goals in an increasingly complex world."

A wide variety of such barriers were identified and discussed in the over 150 letters the review committee received from the community. Collectively, these letters identified several hundred specific problems. We found it convenient to consider them in seven broad categories:

1. Widely varying levels of student ability, and poor preparation for SME&T studies by many.

2. Curricular and pedagogical problems, including a lack of interdisciplinary courses.

3. Ineffective use of instructional technology.

4. A faculty reward system that does not emphasize the importance of instructional effectiveness.

5. The related problem of inadequate use of evaluation for making informed choices about new curricula and teaching methods.

6. Lack of resources for faculty development, for efforts to disseminate improved practices, and to provide modern instructional equipment and materials to their students.

7. Organizational issues: poor institutional articulation *among* institutions (high schools and colleges; two-year and four-year colleges; colleges and employers and states) and *within* institutions (linking teaching and research roles, linking SME&T departments, especially the education and science faculty); resistance to change by key people within academe; indifference to the need for comprehensive change.

The comments we received were often tinged with optimism about the prospects for improvement but also frequently conveyed a sense of urgency. Often there was a satisfaction expressed about the progress achieved over the last few years, intermingled with expressions of alarm about the size of the improvements yet needed.

It is very noteworthy that many contributors to this review used largely the same language to describe both significant improvements in undergraduate SME&T education *and* current barriers. For example, the growing incidence of departments and faculty offering courses that allow students to design their own small-scale research projects and work in teams with faculty guidance to carry these projects forward was noted as an improvement. It was also described as a barrier, namely, it has revealed the poverty of traditional "cookbook" laboratory approaches. What is a significant improvement also has served to reveal and define lingering problems. Frequently, the "problem" or "barrier" side of the "improvement" is the perceived lack of widespread implementation of the "improvement."

What is a significant improvement also has served to reveal and define lingering problems. Frequently, the "problem" side of the "improvement" is the lack of widespread implementation.

The subjects of each of the seven categories are discussed briefly below.

1. Variety in Student Preparation

Participants at the April 1995 convocation on undergraduate education sponsored jointly by the National Research Council and the NSF took note of the tremendous (and growing) diversity of students and of the institutions they attend, and of the concomitant breadth of student needs and institutional objectives [2]. For example, Professor Mary Beth Monroe of Southwest Texas Junior College stated:

> **"I recently heard a speaker comment that although our teaching has improved, learning has decreased overall because our students are not the same students we were nor taught twenty years ago. The conferences, seminars and workshops I have been privileged to attend within the last five years indicate that our educational community is beginning to realize this and is beginning to take important and collaborative strides to address this issue."**

Convocation participants concluded that such breadth would make very difficult the setting of widely applicable goals for what *all* undergraduates should learn and be able to do in science, mathematics, engineering, and technology.

They found the current system, which rarely establishes specific goals beyond those associated with individual classes and courses of study, to be quite unsatisfactory. But, at the K-12 level, the national standards developed for science and mathematics define what students should know and be able to do in these subjects [34]. It was particularly striking to a number of contributors to this review that a need has developed to connect improvements occurring now in K-12 mathematics, science, and technology – and improvements that will be put in place in the future at the K-12 grade levels – with the design and implementation of new undergraduate curricula and improved pedagogical practices. This will require deeper involvement of college and university faculty with SME&T teaching in grades K-12. It will

A need has developed to connect improvements occurring now in K-12 mathematics, science, and technology with the design and implementation of new undergraduate curricula and improved pedagogical practices.

most urgently require attention to evolving K-12 science and mathematics standards by those college faculty actively engaged in improving the preparation of future K-12 teachers, because these teachers will play an important role in increasing the ability of schools to implement improved standards.

Two-year colleges in particular have valuable experience in teaching a student body with widely varying levels of preparation; hence, these colleges could be a significant resource to assist faculty in four-year institutions to work more effectively with students having a wide range of backgrounds.

2. Curricular and Pedagogical Problems

Given these widely varying levels of student preparation, many participants observed that different courses may be required for different kinds of undergraduate students. In many institutions two or more introductory sequences of courses for undergraduates are available. Options range from separate courses rooted in individual departments to interdisciplinary courses that range across SME&T. This diversity of introductory courses is troubling to some. For example, the National Association of Scholars in their recent report [31] noted that in 50 highly selective institutions, the number of courses offered with no prerequisite courses rose from 127 in 1964 to 582 in 1993. It is, however, more sensible to offer single introductory sequences when an institution's lower division undergraduates are relatively homogeneous with respect to education and experience. Once our nation's schools have come closer to meeting agreed science education standards, it should be possible to reduce the number of parallel introductory SME&T courses.

Advanced Placement courses have begun to influence the design of introductory courses in some institutions, according to some faculty with responsibility for freshman courses who wrote or talked to us. There are complicated issues involved. For example, an admissions office can put such emphasis on Advanced Placement course credits from high school that precollege students are discouraged from taking a range of sciences.

The areas of curriculum and pedagogy were identified as rich in recent accomplishments but requiring broader implementation in order to bear full fruit. The National Research Council concluded [2; p.5]:

"Undergraduate education in [science, mathematics, engineering, and technology]. . . is often hampered by outmoded instructional techniques, fragmentation of knowledge by disciplinary specialization more appropriate to advanced research, and frequent inertia in updating and improving undergraduate curricula. [These weaknesses are not inherent in the enterprise. There is under way an explosion of new ideas, technologies, methods for improving the quality of SME&T undergraduate education.]"

Quite a number of letter writers suggested that NSF and other sponsors of educational improvement should turn their attention to the issue of implementation. For example, Rita Colwell, President of AAAS and a distinguished biological scientist, suggested this should be a priority. Joan Girgus of the science program in undergraduate education of the Pew Charitable Trusts wrote:

> **"[T]he most crucial task now facing the NSF and other funders is the conversion of innovation to broad and sweeping change. We know a good deal about what works well for SME&T students; it will require deep commitment to integrate the best of these innovations into the ongoing life of undergraduate SME&T education, thereby effecting the comprehensive educational change that is needed."**

Jerrier Haddad (President, Accreditation Board for Engineering and Technology) put it this way:

> **"Perhaps the largest problem is that piece of academic culture that requires collegial consensus and the leadership of leading faculty and institutions before the bulk of academe will consent to break new ground. Clearly this greatly slows the process of innovation and improvement. It also militates against scholarly research in education."**

There is also the issue of connection between curricula design and the needs of employers. The summary of the 1995 NRC-NSF convocation on undergraduate education states this clearly [2; p.17]:

> **"The needs of the work force are changing (American Society for Engineering Education, 1994; National Academy of Sciences, 1995). Rapid shifts in the labor market are creating a paucity of jobs in some areas and exciting new opportunities in others. This dynamism in the labor market is putting a premium on students who have a broad knowledge of different subjects, skills in synthesizing and communicating information, and the ability to work in teams. Students educated with a narrow disciplinary focus and in solitary learning styles can have difficulties adjusting to such an environment. Indeed, such difficulties are a dominant theme in the complaints voiced by business leaders about contemporary undergraduate education."**

This point was made by a very broad group of letter writers, providing clear evidence that academic administrators and faculty are starting to hear this message, and that scientific societies are taking note, too. For example, in the summer of 1995 the American Institute of Physics (AIP) was completing a survey of students who graduated in physics and were working, in order to rate the importance of various skills and knowledge in their places of employment. Preliminary findings (communicated by AIP President C. Kumar N. Patel) indicated that the top two items were *ability to solve complex problems* and *interpersonal skills*. Leadership skills were also important, and advanced computer skills and technical writing ability were also found to be key in many work settings. However, as indicated by Dr. Eve Menger (Director, Characterization Science and Services, Corning Glass Works, and member of the National Science Board):

"Dynamism in the labor market is putting a premium on students who have a broad knowledge of different subjects, skills in synthesizing and communicating information, and the ability to work in teams."

"In recent years there has been some attempt to remedy this situation but there is a long, long way to go. The topic of realities in the industrial world is usually introduced, if at all, by members of the science faculty who have good intent but little knowledge whereof they speak."

Dr. John McMasters, Principal Engineer of the Boeing Company, described during the November 1, 1995, hearing at NSF [10] a promising program that invites faculty to spend a semester on leave at Boeing facilities

". . . not to show them how we do research, but rather to show them what the graduates of their programs do when they come to work for a company like Boeing. . . We pretty badly underestimated what we could actually accomplish with this."

The problem with a major emphasis on "covering the material" is that many of the facts that constitute this material appear to most students as disconnected nuggets of information.

One important trend reported in the design and delivery of innovative SME&T courses – one that places greater emphasis on concepts and the processes of the disciplines and less emphasis on facts – is considered generally to be a positive antidote to the deadening effects of rapidly and broadly covering a large range of course material. The problem with a major emphasis on "covering the material" is that many of the facts that constitute this material appear to most students as nuggets of information that are both disconnected from themselves and from a context that has meaning and interest to students, such as major societal issues or exploration of new technological opportunities. A valuable way to ground students in the basic processes of SME&T, and perhaps entice them, is to build into introductory courses real or simulated research experiences. The idea is to allow students to experience SME&T as researchers experience it. This design can lead naturally to student discovery of the importance of certain facts. Some who wrote to us also made the point that it is possible to find examples of introductory courses that have gone to excess in avoiding dependence on learning facts. Facts, of course, are essential. It is also possible to find "new" introductory courses that avoid over-emphasis on covering the material, but do not provide the benefit of grounding their students in processes. Innovative courses that succeed for most students seem to have in common a balance between facts and concepts, and they provide – and this is crucially important – a *context* for the material.

3. Ineffective Use of Instructional Technology

One important barrier noted frequently is a specific lack of knowledge about the hardware and technology that has been spreading into increasing use, and to which many students are already attracted. Patricia Cunniff (Director, Science and Technology Resource Center, Prince Georges County Community College [MD]) wrote:

"I see the following as serious problems: (1) successful utilization of the computer in instruction within all types of institutions and within [every] discipline; (2) the challenge of teaching faculty and students how to access, utilize, and incorporate the vast amounts of

> **information available in print and electronically; [and] (3) [learning] how to utilize technology in making education more attractive to students who might otherwise lack motivation or interest in SME&T courses."**

Kenneth Foote (Associate Professor of Geography and Associate Vice President for Research, University of Texas at Austin) observed that:

> **"Senior faculty often lack many of the fundamental skills needed to make use of the information technologies that are changing education. The point has been reached where college students are arriving on campus with information technology skills far superior to three-quarters of the faculty – which is very intimidating to them."**

A very thoughtful letter from Julius Dasch of NASA reported the responses of 143 faculty in the NASA/ASEE (American Society for Engineering Education) Summer Faculty Fellows program and 45 undergraduate or first-year graduate students in summer programs at NASA facilities to the questions posed in Luther Williams' letter. Seventy one percent of the NASA Faculty Fellows stated the biggest problem preventing employment of the best curricular and pedagogical practices was lack of access to hard technology (including software). But, in comparison, 71% of the NASA students nominated poor or apathetic teaching (including faculty woefully unfamiliar with modern technology) as the leading issue, compared to a smaller 38% who directly identified lack of access to modern technology, illustrating that hard technology by itself is not enough to produce improved pedagogy.

"The point has been reached where college students are arriving on campus with information technology skills far superior to three-quarters of the faculty."

Seymour and Hewett [8] observed from their extensive interviews with students and faculty that, often, faculty's first instinct in coping with student dissatisfaction with a course is to modify the course curriculum and seek improvements in instructional technology, rather than tackle the issue of their pedagogical approach, and the appropriate balance among curriculum, technology, and pedagogy.

4 & 5. The Faculty Reward System and Related Problems

The faculty reward system in general was one of the most frequently cited barriers, both by those who wrote to us and in the published literature. The existing balance of rewards is seen to be slanted towards research in part because the system for measuring teaching performance on most campuses does not include broad evaluation of faculty accomplishments in improving the learning of all students. This was one of the three central conclusions of the National Research Council [2; p.5, paraphrased]:

> **Considerable uncertainty surrounds the vital matter of what institutional value is attached to the different kinds of professional work. Faced with this uncertainty, faculty members are apt to stress the one activity for which relatively clear objectives and rewards exist: research that results in peer-reviewed publications . . . Linkages between teaching and research are often neglected. The low priority on teaching is partly a reflection of the low**

priority placed on broadly-defined scholarship. If this were to include a range of design and writing activities related to improving student learning, such as software designed for teaching in innovative ways that promote active student learning, case materials, and even videos aimed at increasing popular understanding of an issue, teaching effectiveness would receive higher priority.

Professor Jaleh Daie (University of Wisconsin; President-elect of American Women in Science) considered the faculty rewards system to be the number one issue:

"Dissonances exist between the research, teaching, and service missions of the university resulting in reduced emphasis on undergraduate education. Furthermore, inconsistencies between a university's mission statement and its policies, procedures, and practices for rewarding faculty create mixed signals. . . Greater integration and a sense of community are prerequisite to effectively engage the faculty in the improvement of undergraduate education. The federal funding agencies share [some] responsibility for such inconsistencies."

Bruce Alberts, President of the National Academy of Sciences, identified this issue as the leading barrier in his letter:

"At major research universities, the most prestigious faculty appointments are those with no teaching responsibilities. The reward structure is especially problematic for junior faculty, for whom earning tenure often depends primarily or exclusively on the number and quality of their research publications. Although many professional organizations and disciplinary societies have published guidelines for recognizing and rewarding scholarly work in education, these guidelines are not widely accepted. [Y]ounger faculty . . . are often advised by their more senior colleagues to delay [developing innovative courses and experimenting with new teaching methods] until after they receive tenure."

We need to inculcate an institutional culture that elevates learning to a position of importance on par with the discovery of new knowledge.

Frequently, discussions of this issue lead to the realization that we need to inculcate an institutional culture that elevates learning to a position of importance on par with the discovery of new knowledge. Progress here would facilitate (among other things) the development of research programs within existing academic structures that focus on learning effectiveness, thereby creating valuable knowledge that can be employed to help make informed choices about new curricula and teaching methods. The lack of adequate rewards for improvements in education is seen by many as the fundamental problem that makes more difficult the widespread acceptance of responsibility for the learning of undergraduates, and puts the burden on single, committed individuals. But, as concluded by the April 1995 convocation [2; p. 5]:

"Undergraduate education will not change in a permanent way through the efforts of Lone Rangers. Change requires ongoing interaction among communities of people and institutions that will reinforce and drive reform."

6. Lack of Resources

This most frequently cited problem is perhaps the largest barrier of them all, because – as many contributors to this review have claimed – it will take substantially more resources to achieve the rapid pace of improvement that is needed. Virtually all of the faculty and administrators from two-year colleges, and many others, who wrote to us stressed three areas in which needs are not being met due to resource scarcity: (1) faculty development; (2) efforts to disseminate good practices among faculty; and (3) the provision of modern instructional equipment and materials to their students.

(1) *Faculty development.* Dr. Dale Ewen (Vice President for Academic and Student Affairs, Parkland College, Champaign, Illinois) wrote:

> "The grave importance of this critical need is evidenced by the following factors: (a) [Nearly] all current faculty have not been trained in the use of the technology and pedagogy [developed in the last few years], (b) The significant use of part-time faculty in most disciplines and their limited access to such professional development is a serious concern/issue, and (c) the very large number of faculty expected to retire within the next five to ten years – will prove a most difficult issue [because] many . . . have no interest in changing their curriculum and teaching styles."

Marilyn Mays, President of the American Mathematical Association of Two-Year Colleges, indicated in her letter to us that part-time faculty usually get left completely out of professional development efforts: "What little professional development money is available in times of reduced resources usually goes to full-time faculty. . ."

The issue of faculty development seems to loom large in the many types of institutions, and across disciplines.

The issue of faculty development seems to loom large in many types of institutions, and across disciplines. For example, Professor C. Kumar N. Patel wrote us (this time as President of the American Physical Society):

> "The American Physical Society believes our faculty are still not, in general, familiar with the research about learning and the positive impact of research-based learning strategies on student performance. Most of the innovation occurs in the classes and laboratories of individual professors and has little impact on other faculty at the same institutions. In general there seems to be greater effort needed to support institutional systemic change."

Jean MacGregor (Interim Director of the Washington Center for Improving the Quality of Undergraduate Education, Evergreen State College) reflected a similar sentiment in her letter, observing:

> "The biggest problem still facing us in our efforts to improve SME&T education is lack of adequate faculty development. There is a crying need . . . for faculty members to have the time and support they need to adapt reform curricula to their own teaching contexts."

(2) *Dissemination of good practices.* A number of contributors to this review emphasized the need for easy access by faculty to databases in

which vital information about new courses and curricula could be explored. This is one of the major ingredients necessary for faculty development. C. Bradley Moore (Professor of Chemistry at the University of California, Berkeley, and Chair of the National Academy of Sciences Committee on Undergraduate Science Education) wrote that there is a need for materials specialized to all types of institutions, reflecting the importance of the diverse curricula offered by American undergraduate institutions to a wide variety of students, and suggested that these materials should be available in "databases [designed] for faculty to find appropriate materials for individual courses."

Jane Coulter, Deputy Administrator, Science and Education Resources Development, of the US Department of Agriculture's Cooperative State Extension, Education, and Extension Service, nominated the need for dissemination of knowledge as a key issue for USDA:

"We truly believe and can provide evidence that USDA's higher education competitive grant programs are, indeed, addressing regional, national, and international educational needs in the SME&T areas. However, so much more could be gained if better mechanisms were in place for the dissemination of project products and results. Dissemination of outstanding program/project accomplishments needs to be enhanced."

To underscore this point, the need for a dissemination link between the development of innovative approaches to undergraduate instruction, and the implementation of these innovations emerged as a key point from the 1995 joint NRC-NSF Convocation *From Analysis to Action* [2; pp.5-6]:

"[I]nnovations and successes in education need to spread with the speed and efficiency of new research results."

It seems clear also that a better dissemination system that incorporates assessment and validation would be of significant assistance in changing the faculty reward system. Research may have such a prominent role in the current faculty reward system in part because of the well-established peer review and dissemination mechanisms that are in place.

(3) *Provision of modern instructional equipment and materials.* Good pedagogical practice is facilitated by access to modern instructional equipment that is not widely available in American institutions. C. Peter Magrath (President of the National Association of State Universities and Land Grant Colleges) indicated in his letter that:

"Today, there is a serious decline in the quality of academic facilities, particularly in terms of state-of-the-art instructional materials and equipment, high-performance computer technology, and well-equipped instructional labs and centers."

This sentiment was echoed widely in the letters we received, and during the 1995 convocation. A straightforward framing of this issue as it applies broadly to the teaching of introductory courses was stated by Jerome Pine, Professor of Biophysics at California Institute of Technology:

> **"There is the problem of administrative support for non-lecture introductory courses aimed at the entire college community. They may cost more in personnel, space, and materials. It must be realized that there is no free lunch, and that quality science instruction will cost more than mass-produced lecturing. More need not mean prohibitively more, however. It simply means more than now, which is too little. Computer science also costs more than it used to, and it gets the support it needs."**

7. Organizational Issues

One of the most imposing barriers identified by our correspondents was the collection of rigid organizational structures and poor attention to the linkages between and among various types of organizations – which many refer to as "articulation issues." One of the most frequently mentioned issues was the very high autonomy of individual faculty in departments, and also the autonomy of departments themselves. Dr. Joseph Perpich (Vice President for Grants and Special Programs, Howard Hughes Medical Institute) considered this to be the foremost barrier, writing:

> **"Reform of departmental structures is needed to foster educational collaboration among faculty members from different scientific fields."**

The importance of a sense of departmental responsibility is indicated by the work of Sheila Tobias [35]. Her studies show that the institutions at which things work have departmentally-based continuous improvement philosophies rather than relying on outside grants that expire (after which the situation reverts to what it was previously).

The institutions at which things work have departmentally-based continuous improvement philosophies.

There are numerous types of organizational connections that need further development. Frequently cited was the need for improved articulation between two-year and four-year institutions - to ease the transition of students from the former to the latter, and to ease the burden of paying for college by allowing undergraduates to take fuller advantage of community colleges. This requires that much collaborative work be done by faculty in both types of institutions.

A similar issue is the need for collaboration between faculty in schools or colleges of education and the SME&T faculty. Numerous contributors to our review identified the need for improved preparation of undergraduate teachers; a subset targeted joint work between education and SME&T faculty as the key means of accomplishing this. Lack of interaction among SME&T faculty, faculty in other academic disciplines, and faculty in schools of education is a serious flaw in much of precollege teacher preparation, arising in a separation of

Lack of interaction among SME&T faculty, faculty in other academic disciplines, and faculty in schools of education is a serious flaw in much of precollege teacher preparation, arising in a separation of methods from content.

methods from content. The National Research Council [2] recommended that content should *not* be separated from methods in preparing future teachers; the two should be embodied in the same course or run in parallel – desirable outcomes that require interdepartmental interaction.

Another frequently cited issue was the need for more joint activity between academic institutions and employers. This issue was most frequently mentioned by employers who wrote to us or who testified at our hearings. However, it was also recognized as an issue of growing importance in other studies [16] and was mentioned a number of times by academic administrators. The program described by John McMasters (page 44) is another good example of this interest in increased joint activity between academic institutions and employers.

Lowering the Barriers and Meeting New Expectations

When all of these pieces of the picture of the situation today are put together, it is apparent that there have been major changes in the world, the economy, and higher education, creating new expectations for undergraduate SME&T education, different from the situation of 10 years ago, when the major focus was on the "pipeline problem" – the production of an adequate number of professionals in the individual SME&T fields, resulting in a focus on those majoring in SME&T. These new expectations are reflected in the broad charge to this review committee and in the statement of our vision. It is also clear that there are major barriers to meeting these new expectations.

A modest investment in NSF education programs has leveraged significant improvements on many campuses.

During the last ten years, we have seen NSF programs lead to significant improvements. Innovative ideas of individual faculty members, in various departments on many campuses, were funded through these programs. These *single changes* evolved into broader curriculum projects such as those referred to as "the Calculus Reform movement" and to consortial efforts such as the *NSF Collaboratives for Excellence in Teacher Preparation* (CETP) and the *Alliances for Minority Participation* (AMP). Those broader efforts have now led to the beginnings of institution-wide reform on college and university campuses. A modest investment in NSF education programs has, therefore, leveraged significant improvements on many campuses. But, as we have seen from the community's description of barriers, most of the changes are still rather isolated and transitory – too often dependent on a faculty member acting *in solo* (a "Lone Ranger") – and not well disseminated, even to other parts of the host institution. Too few of the nation's undergraduates are affected. Systemic changes that will spread and sustain reform have been only tentative.

During the October 23, 1995, hearing [*Disciplinary Perspectives*] Professor Alan Tucker of SUNY Stony Brook summarized letters written as part of our review by members of the mathematics community: "[T]omorrow's calculus student will have a better understanding of the power of math than the student today. This is something that you folks at NSF deserve credit for. Lots of little things have been tried over the years, but to break out of a rut, one really needed a little extra help, and one needed the confidence of this funding to believe it was worth trying and pushing further. It was an initiative that in the end involved close to 200 awards . . . and this doesn't include subcontracts, which would double the number here . . . I just think the impact of this was a grassroots effort, and this is the strength of higher education, that there are so many good people out there waiting to be helped. Hundreds and hundreds of workshops were run. . . typically six or seven faculty had been to reform workshops before they had started changing. Now there are hundreds of pedagogical talks at the national meeting . . . NSF did this with $3 million dollars a year for six years. Calculus as an industry is over a half a billion dollar a year business. So this was a fraction of one percent for just half a decade, and yet it has changed things."

The specific recommendations we make in the next section reflect the conviction, based on all that we have heard, read, and learned during the course of this review, that what is necessary is that undergraduate SME&T education become:

- **More central** in the curriculum, in faculty rewards.
- **More centered** in the student, in the *processes* of the disciplines.
- **More connected** with student experience, across knowledge boundaries.
- **More collaborative** among students, across institutions/ organizations/ industry.
- **More comprehensive** across each institution, in all institutions.

To produce increased learning in all students, the

SME&T undergraduate education community will have to broaden its view so as to

- *seek to serve **all** students* – focusing not on what faculty teach but on what it expects those students to learn; and
- *develop more effective curricula and pedagogy* – drawing on research knowledge about human learning and on technology.

> "It dawned on me about two weeks into the first year that it was not teaching that was taking place in the classroom, but learning."
>
> — Pop star Sting,
> reflecting on his early career as a teacher

Institutions must systemically support these efforts, both

- *internally on the campus* (e.g., helping graduate students learn how to facilitate student learning as part of their training; advancing research on human learning and its use in practice; reallocating funds to provide for faculty development and instructional equipment; working to break down walls between departments; focusing faculty reward systems more on teaching and learning; holding whole units accountable for enhanced student learning); and

- *through links with other organizations* (such as articulation agreements to facilitate student transfer and partnerships to enhance teacher preparation programs).

Funders and other external constituents must encourage and enable these changes to occur throughout the array of institutions providing undergraduate SME&T education, while holding those institutions accountable for results. Our recommendations are designed to accomplish these changes.

Before detailing specific recommendations, it may be helpful to the reader to describe our conclusions more generally and provide references to the listing of recommendations in the next section (pages 61-72), where they are grouped according to the constituency to which they are directed. As indicated in the previous paragraph, our review leads us

TO ASK the SME&T undergraduate education community to:

- *seek to serve* all *students*, focusing on student learning. (See recs. I.B; IV.A; V.B; VI.A, B, F-H; VII.A, B, E, F, I, J.) Study in SME&T fields must no longer be seen solely as narrow preparation for particular, specialized careers, but must be accepted as important to *every* student, including those groups historically underrepresented in SME&T. We recognize that there are vast differences in basic ability levels across humankind; not all students will be able to learn equal amounts with equal effectiveness. But all students must have an opportunity to learn and decades of research and experience have demonstrated that if one *expects* a pupil to learn, learning likely *will* improve, often substantially [36]. In particular, the SME&T undergraduate education community will have to be concerned much more inclusively about:

a) *Preparation of K-12 teachers in these fields*. (See recs. II.C; IV.D; VI.C; VII.H, I.) Programs to prepare teachers must include substantially more scientific and mathematical content, with stress on agreed standards and on the methods and processes of science. In particular, prospective teachers must understand that even "hands on" laboratory experiences may or may not be inquiry-based (i.e., involving the student in asking questions and finding answers). Teacher preparation programs must also have strong components in areas such as the arts and humanities, recognizing the vital role these fields play in supporting teachers in their responsibility to foster creativity, curiosity, and inquiry.

Prospective teachers must be expected not only to learn SME&T content but also to come to appreciate that ***all*** students can learn and to approach their tasks as teachers from that perspective. They must also learn how to become effective facilitators of student learning in their teaching careers. There is increasing evidence that prospective teachers can learn about teaching mathematics from studying the "practice" of mathematics teaching. A recent letter report of the Mathematical Sciences Education Board cites a number of ways in which even substitutes for classroom expertise (e.g., videos or case studies) can assist in the preparation of teachers. Actual experience in a classroom with a master teacher and involvement with prospective employers of high school graduates, however, are extremely important parts of an effective program to prepare teachers.

Since an increasing number of people seek to enter K-12 teaching some time after having completed basic courses in SME&T at the collegiate level, special upper division courses in SME&T that stress inquiry and reflect sound pedagogy are becoming essential, as the previous SME&T coursework taken by the student in many cases will not have met the expectations of these recommendations.

A large percentage of prospective K-12 teachers begin their education in two-year colleges. These institutions, with their clear commitment to teaching and with so many prospective teachers as students, must be more significant partners in the system of teacher preparation.

Two-year colleges must be more significant partners in the system of teacher preparation.

Regardless of the specific structure and content of teacher preparation programs, improved undergraduate SME&T teaching generally is an essential precondition of more effective teacher preparation. It seems likely that elementary and secondary teachers will tend to teach as they were taught as undergraduates.

It should be kept in mind also that many prospective teachers prepared at the undergraduate level eventually will become principals and superintendents. In such leadership positions, they will influence the learning throughout entire schools and school systems. Their attitudes toward science, mathematics, and technology, and ultimately the extent to which they can improve education in those fields, will depend in part on the kind of experience they had in SME&T as undergraduates.

It surely is a function of higher education to provide a basis for meaningful employment.

b) *Needs of persons going into the technical work force.* (See recs. I.B; II.B; VI.E; VII.D, J.) We have discussed at length the changes occurring in America's work force and the perceptions of employers about our graduates. SME&T education has value in and of itself, but it surely is a function of higher education (not to mention an expectation of students and parents) that a college education will provide a basis for meaningful employment.

SME&T programs must include appropriate provision for the short-term needs for many kinds of skilled positions requiring an associate-level education; but such programs should incorporate basic principles of SME&T as well as particular vocational skills, so they do not become useless in a brief period. Programs must have strong linkages with employers, with adequate opportunities for internships and practicum experiences (note, in this regard, the success of programs like *School-to-Work* and *Tech-Prep*).

At the baccalaureate level, SME&T programs should analyze the needs of employers and work with them to ensure effectiveness of preparation. At any level, programs must recognize that employers need graduates who can communicate and who can work in teams to solve problems; such skills, along with lifelong-learning skills, for example, do not simply appear mysteriously in students.

Do our programs for majors really provide students with a basic understanding of SME&T generally, and acquaint them with the methods, processes, history, and societal context of SME&T?

c) *Preparation of majors in these areas.* (See recs. VI.B, J; VII.C, G, J.) SME&T does a generally good job of preparing majors, particularly those headed for graduate schools and eventual academic employment. Even here, however, there is work to be done – as suggested by the following questions: What is really essential for majors to understand and be able to do? How do our programs prepare them to work at the boundaries between fields, where much cutting-edge work goes on? Do our programs for majors really provide students with a basic understanding of SME&T generally, and acquaint them with the methods, processes, history, and societal context of SME&T? Do we adequately help our majors understand the ethical responsibilities of SME&T practitioners? Should we do more to help those going to graduate school prepare for the different environment and culture of a doctoral program, for example? Do we adequately provide our students information about career options for majors? And how can our programs better enable students to acquire the ability and motivation to continue learning and the skills to communicate and work effectively in teams?

d) *Science literacy for all.* (See recs. VI.A, B; VII.A, C, E.) This is a difficult topic, as there is not substantial agreement about what is meant by "science literacy" or whether it is actually feasible for *all* students to acquire it. We believe that what *is* necessary is that every student have an opportunity to learn what science (broadly defined) actually is; what SME&T professionals actually do; how to evaluate information presented as "scientific"; and how society should make informed judgments about science and engineering. Not only is such knowledge important for the exercise of good citizenship in a technical world, but an educational system that provides this opportunity also gives individual students a basis for further work in SME&T fields at some later date – it expands options rather than closing them off.

- *develop more effective curricula and pedagogy.* In addition to the curricular and pedagogical suggestions contained in a), b), c), and d) above, there are some general principles that are clear from our review. (See recs. IV.A; V.M; VI.B-E, G-I, K; VII.A-F, I.)

The variety of student preparation requires a variety of pedagogical methods. The effective use of technology and the modularization of courses can also assist faculty to deal with individual differences

The variety of student preparation requires a variety of pedagogical methods. **Technology and modularization can also assist faculty to deal with individual differences.**

It is certainly **not** the case that students enter higher education knowing exactly what they plan to do and then sticking with that plan. *There should*, therefore, *be in place a curriculum as flexible as possible.* That curriculum should provide reasonable options for students to move onto or off of different career-preparation paths (e.g., a), b), and c) above) as well as the useful understanding of SME&T fields needed by all students for the effective exercise of citizenship. Adequate advice should also be available to students about the various career opportunities for those prepared in SME&T fields. Thought also should be given to pathways through which students might gain access more easily to science and mathematics; for instance, study in engineering, technology, the earth or the social sciences may provide easier entry for more students to science and mathematics than more traditional pathways.

"An appreciation of what is happening in science today, and of how great a distance lies ahead for exploring, ought to be one of the rewards of a liberal arts education. It ought to be a good in itself, not something to be acquired on the way to a professional career but part of the cast of thought needed for getting into the kind of century that is now just down the road. . . It is worth a try." Lewis Thomas [37].

There must be high expectations of every student, but there must also be a welcoming and encouraging climate for learning. A hostile atmosphere assuming that only a select few can learn at a worthwhile level does not produce good results for the many - and leads many students to leave SME&T entirely. For faculty, their disciplines are exciting and fun; SME&T education should convey that same sense to students. SME&T education must connect desired learning with the student's previous experience and with other fields in which the student is learning, incorporating interdisciplinary work that transcends even SME&T departments. It must include attention to skills and attitudes that students must learn (e.g., the ability to work in teams to solve problems and the ability to communicate with others). SME&T education must put greater emphasis on active, collaborative learning; focus on the processes of inquiry and discovery; and rekindle the unique curiosity, the sense of wonder, with which every child is born. Consider the following passage from *The Washington Post* by columnist Steve Twomey about his son Nick's first birthday (March 25, 1996):

"My son tries to pick up holes. . . He tries to pick up shadows, too. But I'm being unfair. There is nothing he won't try to pick up, because there is no such thing as an uninteresting object, and I'm really kind of jealous. Nick has a full sense of wonder, and I don't anymore."

TO ASK institutions to support these changes systemically:

- *internally on the campus*. The evidence from this review is clear – the improvements achieved have not been widely implemented and are not sustainable without significant change in the culture, policies, and practices of higher education. It is not enough that individual faculty members in isolated ways advance student learning. Many contributors to this report have suggested that what we need is not *more innovation* but *more implementation*, so that local improvements are both spread throughout the institution and made sustainable over time. Otherwise, gains will be transitory and depend on the comings and goings of individual faculty and administrators.

What is needed now is institutional action, not just commitment. Especially noteworthy is the need to modify departmental and faculty reward structures and policies, as these will determine where faculty devote their energies.

What is needed now is institutional action, not just commitment. It does little good, for example, for administrators to speak eloquently about the importance of the undergraduate experience while setting budgets based on a department's ability to attract research dollars. What is important is that every institution ensure that its mission; its personnel, planning, and budgeting decisions; student support mechanisms; and facility designs all support enhanced undergraduate learning in SME&T. Especially noteworthy is the need to modify departmental and faculty reward structures and policies, as these will determine where faculty devote their energies. For instance, promotion and tenure policies can discourage attention to undergraduate teaching and learning, and the definition of "research and scholarship" can leave out the kind of broad scholarship so powerfully urged by Ernest Boyer. Faculty should be more engaged in the intellectually exciting questions about how students learn; institutions should foster that engagement and support research, both discipline-specific and discipline-independent, about human learning. (See recs IV.B; V.A-J; VI.J.)

Resource allocations are equally important – including funds for equipment, decent facilities, and classroom and laboratory assistance. It is perhaps even more essential, however, that institutions provide for comprehensive faculty development and for faculty time to consider, plan, and implement desirable changes. Institutions offering graduate programs should see that preparation for careers as effective undergraduate teachers is included in those graduate programs – to aid in increasing the number of faculty committed to undergraduate education and able to help students learn more effectively. Similarly, institutions of higher education should make a commitment to more effective programs of preparing K-12 teachers, not only as a contribution to education generally but, as an investment, to improve over the long term the preparation of future incoming first-year students and so diminish the need for remedial education. (See recs. IV.B; V.E, I; VI.J.)

It is essential that institutions provide for comprehensive faculty development and for faculty time to consider, plan, and implement desirable changes.

Institutions should also act to reduce organizational rigidities by fostering cooperative efforts and reducing financial disincentives to interdepartmental work. *Walls between departments can be broken down.* (See recs. V.I, L, M; VII.G.)

- *through links with other organizations.* As has been repeatedly noted by participants in this review, no single part of the education continuum can, on its own, accomplish what needs to be done. Institutions of higher education must work together with the schools and with employers to lift undergraduate SME&T education to the next level. We do not suggest these linkages in order to create the *appearance* of cooperation; not only are meaningful partnerships *necessary* to make education easier for students, but numerous examples in the body of this report illustrate the ways in which different organizations have specialized knowledge and/or experience to offer, to the benefit of the whole. (See recs. II.C; IV.C; V.I,K; VII.H.)

Institutions of higher education must work together with the schools and with employers to lift undergraduate SME&T education to the next level.

TO ASK funders and other external constituents to enable and encourage these changes while holding institutions accountable.

The context in which SME&T exists must be supportive; namely, it is vital that the organizations and agencies that relate to higher education act thoughtfully to reinforce and accelerate the kinds of changes we recommend. It may be time, for example, for a new definition of the relationship between higher education and the Federal government.

In the history of this nation, such Federal initiatives as the various Morrill Acts and the GI Bill have substantially changed the course of higher education. Should not there now be established a new framework, through legislation, that would recognize the essential role that SME&T and technology transfer play in the national well-being; that would reconnect the research base of our institutions to both the learning of students and the wider society; and that would provide incentives for implementing systemic reform of SME&T undergraduate education in America's two- and four-year colleges and universities? (See recs. I.A-C; II.A-C; III.A-C; IV.A-E; V.F.)

It is essential that criteria for accreditation focus on teaching effectiveness and learning outcomes.

It is essential that criteria for accreditation focus on teaching effectiveness and learning outcomes rather than on formulaic counting of courses; some old models reflect attitudes of "turf protection" that reinforce the kinds of interdepartmental and inter-institutional barriers that good practice seeks to lower. An accrediting agency should pay attention to inter-institutional relationships and the movement of students between institutions (examining the facility of transitions from K-12 education to the collegiate, from two-year colleges to four-year, and from undergraduate to graduate or professional); it should look for evidence of active partnerships with K-12 and with business and industry; examine programs to help both new and seasoned faculty be more effective teachers; and consider carefully the faculty reward system. (See recs. VII.A-J; VIII.A, B.)

Karl Pister, Chancellor of the University of California at Santa Cruz, wrote in a recent article [38]:

"Through the structure provided by the federal land-grant college legislation in the nineteenth century, the nation's public universities were instrumental in the successful development of natural, industrial, and agricultural resources. By updating the land grant model, we can put universities to work developing our vastly underutilized human resources - and, in the process, tackling the immense challenges we face today."

"Three cultural shifts must occur if we are to succeed. First, we need to encourage innovative ways of looking at problems, moving away from the increasing specialization of academia to develop new interdisciplinary fields that can address complex real-world problems from new perspectives. Second, the orientation of faculty effort and the faculty reward system in our universities must support the full range of institutional missions in a more balanced manner. Third, our society must be willing to make quality education, especially in science and technology, accessible at all levels for all students. Education must be seen more as an investment in society's well-being and less as a cost."

Professional societies, publishers, and testing agencies all must support institutions working to implement more effective SME&T education. Flexible teaching materials must help students learn attitudes and processes as well as facts. Good curricular resources must be available to teacher educators in science and mathematics. Assessment tools must reinforce departmental and faculty goals for student learning. Professional societies must provide forums for research results in learning and for dissemination and adoption of sound educational practice. (See recs. IX.A-D; X.A-B; XI.A-B.)

The NSF itself must reflect through its organization, its rhetoric, and its leadership activities, a holistic approach to undergraduate SME&T education. Its research directorates must continue their good work to integrate research and education, while the Directorate for Education and Human Resources continues to provide overall leadership through a cadre of scientists, mathematicians, and engineers *whose first priority is education* and who provide a multi-disciplinary view not available elsewhere. Both the disciplinary perspectives of the research directorates, focused primarily on research, and the student-centered perspective of EHR, focused primarily on education, are necessary – since research and education are neither divorced nor identical. (See recs. XIII.A.3, A.4, A.6, A.9, C.2, C.4.)

Programs of the National Science Foundation and other funders intended to improve undergraduate SME&T education *must* be designed to provide explicit incentives to make the kinds of educational changes we are urging on institutions, departments, and faculty. There must be a variety of such programs, since it is not the case that "one size fits all"; but these programs must have attributes in common of consistency with our recommendations to the SME&T community, of sustainability, of validation and dissemination, and of propagation to other settings.

Programs of the National Science Foundation and other funders *must* be designed to provide explicit incentives to make the kinds of educational changes we are urging.

There must be continued and expanded close collaboration among Federal agencies, industry groups, and foundations so that their programs complement each other and do not send mixed messages to the SME&T community. Programs must be sustained over a period of years, since educational improvement is a long-term process; and the NSF should do all it can to encourage creation of other funding sources to sustain educational enhancement, e.g., endowments at the statewide or institutional level. Evaluation of programs and assessment of results must also be long-term; for instance, the NSF should return to a campus long after expiration of a grant to determine the permanence of desired changes, and it should fund longitudinal studies of performance of graduates over many years. All NSF educational programs must stress activities that have considerable direct impact on student learning rather than those that focus on creation of administrative structures. (See recs. XII.A-B; XIII.A.1, A.5, A.7, A.8.)

Finally, there must be expanded funding of NSF undergraduate programs to provide leverage to accomplish all these changes. The Neal Report focused on SME&T majors only, but even its recommendations have not been fully funded. In fact, as pointed out

NSF's funding for undergraduate education must be at least doubled in real dollars in the next decade if this nation is to have any hope of accomplishing what is described above.

earlier (see pages 15-16), there is a gap of more than $73-million between present funding of three key instruction-related programs (Laboratory Development and Instructional Instrumentation; instruction-oriented Faculty Professional Enhancement; and Comprehensive Improvement Projects) recommended by the Neal Report and the level of support which it urged. To overcome just that gap would require a 41% increase in present funding of NSF undergraduate programs. That present funding, however, reaches only some of the students in institutions enrolling fewer than half of America's undergraduates (see page 33), and its good results have not been deeply and widely implemented across those institutions. We are convinced, therefore, that the NSF's funding for undergraduate education – and hence its ability to leverage much greater sums – must be at least doubled in real dollars in the next decade if this nation is to have any hope of accomplishing what is described above. This increased level of support for undergraduate education is necessary, in part, to sustain the reform of K-12 education, which has been a high priority for the nation and to which substantial resources have been allocated. Such an increase for undergraduate education programs at the NSF would amount to approximately $178-million, an amount representing only about 0.8% of the total estimated expenditures by higher education institutions on undergraduate SME&T education. We believe we cannot afford to do less. (See recs. XIII.A.2, B, C.1, C.3.)

Our recommendations, therefore, are comprehensive, designed to make possible continuous improvements in undergraduate SME&T education so as to maintain, in an ever-changing environment, a level of quality that will ensure economic security and cultural richness of life for each of America's citizens, for

we can no longer be satisfied with incremental improvement in a world of exponential change.

IV. Recommendations

To accomplish the ambitious goals we have set forth will require actions of various kinds by many different individuals and agencies working in concert. Our recommendations to these several audiences are listed below. We hope that *no one* will read these recommendations and assume that *someone else* will carry them out - everyone has an important part.

Our recommendations are not intended to be prescriptive, but rather a framework for improved undergraduate SME&T education within which we hope individuals will act and institutions and organizations will make decisions and allocate resources in ways specific to them. While all of the recommendations are important, the National Science Foundation bears special responsibility in this area: its leadership and funding of programs to improve undergraduate SME&T education have been the drivers of the significant progress made in the last ten years, and will be central to the even greater progress we must make in the next ten.

The interrelatedness of our recommendations should be noted carefully. It is crucial to successful implementation of these recommendations that everyone concerned with undergraduate SME&T education, from policy-makers to administrators and teachers, *work in concert* – to the benefit of student learning and ultimately, therefore, to the benefit of this society.

We recommend that:

I. The President and the Congress

A. Establish, in consultation with the higher education community, a new social contract for higher education in America. What is needed may be a new *act* to reconnect the research base of these institutions to the learning of students and to service to the wider community.

B. Develop a national human resources policy that focuses on using the talents of *all* of America's people to improve the nation's productivity and quality of life.

C. Support a level of federal funding for undergraduate SME&T education sufficient to accomplish the recommendations of this report, recognizing that the undergraduate level plays a critical role in the continuum of educational programs and that SME&T fields constitute a cornerstone of America's future economic and political security.

II. Business, industry and other employers of SME&T undergraduate students

A. Participate in national and state public policy decisions, helping those making such decisions (as well as their own colleagues) understand the critical importance of quality SME&T education to the national well-being.

B. Communicate clearly to colleges and universities their expectations about the desired characteristics of potential employees and, hence, of the kind of educational preparation needed at the undergraduate level.

C. Expand both partnerships with – and funding of – colleges and universities to advance institution-wide reform initiatives, expand student internships and faculty exchanges, develop more practicum experiences for students, make more effective use of existing resources, and provide for the retraining and continuing education of the work force (including K-12 teachers).

III. National and regional media

A. Become better informed about the broad range of issues in collegiate-level science, mathematics, engineering, and technology education.

B. Inform the public more fully about the issues and successes in undergraduate SME&T education and its critical significance to the future well-being of this nation.

C. Continue and expand engaging ways to bring SME&T content to the general public, to increase awareness that will foster a context of concern for education in these fields.

IV. State governments and statewide higher education boards

A. Raise expectations about the kind and quality of undergraduate SME&T learning in the state.

B. Ensure that funding formulas and state policies are modified, as necessary, to provide incentives and rewards for increased undergraduate student learning in SME&T at institutions in the state.

C. Encourage formal and informal collaborations between institutions, especially between two-year and four-year colleges, including development of firm articulation agreements to make student transitions efficient and effective.

D. Collaborate with external accrediting agencies to make strengthened science, mathematics, and technology standards for K-12 the norm in accrediting teacher education programs and in licensing K-12 teachers.

E. Raise, over time, expectations for achievement of students graduating from high schools so as to reduce demand for remediation at the postsecondary level.

V. College and university governing boards and administrators

A. Reexamine institutional missions in light of needs in undergraduate SME&T education.

B. Accept responsibility for the learning of all students and make that clear not only by what the institution says but also by putting in place mechanisms to discharge that responsibility at the institutional and departmental levels.

C. Hold accountable and develop reward systems for departments and programs, not just individuals, so that the entire group feels responsible for effective SME&T learning for all students.

D. Ensure that in hiring faculty careful attention is paid to interest in teaching, ability to help students learn, and concern for students.

E. Provide resources to ensure that faculty, particularly new faculty, have the opportunity to both learn how to and have the time to design effective instruction, use technology appropriately, foster inquiry-based and collaborative learning, and assess learning achieved.

F. Inform the public and the media about what SME&T professionals do and be advocates for the importance of SME&T to society.

G. Make sure that the faculty reward system, in practice as well as in theory, supports faculty who effectively help students learn in hospitable environments that recognize individual student differences and that provide reasonable opportunities to address those differences.

H. Ensure that there is a supportive climate across the campus for student learning, including sound academic advising and effective career development services (so that the varied opportunities available to persons having backgrounds in SME&T are known to students).

I. Create or strengthen an institution-wide commitment to the preparation of K-12 teachers and principals, bringing together departments of education, SME&T and other departments, K-12 staff, and employers of teachers to design and implement improved teacher preparation programs having substantial SME&T content and stressing rigorous standards, along with emphasis on engaging students in learning.

J. Support broad-based, sound, peer-reviewed research on human learning (both discipline-specific and more general), so more will become known about how undergraduates learn most effectively, and encourage cross-campus conversation about this topic.

K. Reach out in partnerships with other institutions of higher education, the schools, informal science education organizations, and employers to improve SME&T education collaboratively.

L. Ensure that responsibility for lower level SME&T courses resides in the full-time permanent faculty.

M. Encourage and support development of interdisciplinary SME&T curricula.

VI. SME&T departments

A. In collaboration with other departments and with prospective employers, set departmental goals for undergraduate learning. These goals must include clear expectations, the attainment of which is measurable, about what all students in the institution should learn.

B. Provide a curriculum that engages and motivates the broadest spectrum of students, enabling every student to learn and providing reasonable flexibility for students to move onto or off of various career-preparation paths without undue penalty.

C. Develop upper division SME&T courses and other educational opportunities appropriate for prospective and current K-12 and two-year college faculty.

D. Foster interdisciplinary education.

E. Develop meaningful connections with employers to provide appropriately responsive educational experiences for prospective and current members of the work force.

F. Encourage faculty to work toward the understanding of and resolution of serious educational issues, and reward those who most effectively help all students learn.

G. Create and support learning communities for students and faculty, including clubs, social events, and peer learning and group study opportunities.

H. Use instructional technology effectively.

I. Make use of resources available in colleges and departments of education to strengthen the pedagogical foundations of SME&T undergraduate education.

J. Provide opportunities for graduate students to learn about effective teaching strategies as part of their graduate programs.

K. Encourage and participate in research on learning.

VII. SME&T faculty

A. Believe and affirm that every student can learn; recognize that different students may learn in different ways and with differing levels of ability; and create an environment in each class that both challenges and supports.

B. Be familiar with and use the results of professional scholarship on learning and teaching.

C. Build into every course inquiry, the processes of science (or mathematics or engineering), a knowledge of what SME&T practitioners do, and the excitement of cutting-edge research.

D. Devise and use pedagogy that develops skills for communication, teamwork, critical thinking, and lifelong learning in each student.

E. Make methods of assessing student performance consistent with the goals and content of the course.

F. Start with the student's experience; understand that the student may come with significantly incorrect notions; and relate the subject matter to things the student already knows.

G. Build bridges to other departments, seeking ways to reinforce and integrate learning, rather than maintaining artificial barriers.

H. Develop partnerships and collaborations with colleagues in education, in the K-12 sector, and in the business world, to improve the preparation of teachers and principals.

I. Model good practices that increase student learning.

J. Take seriously academic advising that helps students have as much flexibility as possible and is linked to career development services of the institution.

VIII. Accrediting agencies

A. Incorporate the principles of effective SME&T education for all students (as identified in V, VI, and VII above) into criteria for institutional accreditation.

B. Look for evidence in student learning outcomes of genuine commitment to and effective practice of undergraduate education that reaches all students, including groups historically underrepresented in SME&T.

IX. SME&T professional societies

A. Work together to promote *education* as well as *research*, focus on *student learning* as well as *teaching*, and help departments in their disciplines find realistic ways to implement these recommendations.

B. Ensure that their activities, especially their journals and meeting programs, reflect concern for SME&T education for all students.

C. Assist members to obtain working knowledge of research results in undergraduate education.

D. Encourage and support participation of undergraduates in their regional and national meetings by providing opportunities for undergraduate research symposia as well as including on programs some topics of interest to undergraduates.

X. Producers of educational materials

A. Work closely with creative faculty to develop and refine the high-quality, flexible instructional materials required to support the educational environment described in previous sections.

B. Disseminate these materials, assist faculty in their effective use, and provide feedback to authors.

XI. State and national testing agencies

A. Develop, in collaboration with the higher education community, assessment tools that better measure achievement of the kinds of SME&T learning recommended herein.

B. Support research that will aid in the validation of the extent to which these tools actually measure what is intended.

XII. Private foundations and Federal funding agencies

A. Recognize the interrelatedness of reform efforts; support a common agenda for improving undergraduate education through programs that provide long-term, consistent attention to undergraduate education; and focus resources on strategic investments that will compliment the support of others and lead to wide-spread reforms.

B. Participate, in particular, in the promotion and support of institution-wide reform at institutions making a commitment to systemic change.

XIII. The National Science Foundation

As an essential part of its dual mission of education and research, the NSF must make clear the high priority of undergraduate education. To do so, it is crucial to have within NSF a unit (the Division of Undergraduate Education, within EHR), staffed by practicing scientists, mathematicians, engineers, and technologists from many disciplines, that has undergraduate education as its *first* priority and that relates to *all* institutions providing undergraduate SME&T education. Such a unit is critical to maintain strong linkages with NSF's discipline-oriented research directorates, which must continue to support undergraduate education within their specific fields. This framework, consistent with the view that education and research cannot be divorced, will enable NSF to act effectively in three arenas (A, B, C below) to lead and support the improvements in undergraduate SME&T education reflected in the preceding sections of these recommendations.

A. Take actions internally to NSF that will advance undergraduate education:

1. Lead the development of a common national agenda for improving undergraduate SME&T education, in a collaborative way with other Federal agencies and foundations (see XII. A. above).

2. Strengthen its efforts to engage all two-year colleges, four-year colleges, and universities (and their faculties and, as appropriate, their students), as well as informal science organizations, in all of its education programs.

3. Make clear to all colleges, universities, and other educational institutions receiving grants and contracts, that the NSF expects its awards to contribute positively to the quality of undergraduate SME&T education.

4. Establish an intra-Foundation mechanism to provide a forum for discussion of undergraduate SME&T education issues and to establish a common agenda and an NSF-wide set of strategies for improving undergraduate education.

5. Lead the development of a rigorous research agenda about human learning at the undergraduate level to answer questions not yet fully answered about SME&T education, ensuring that results flow back into funding priorities, program decisions, and educational practice.

6. Ensure that good principles of SME&T education are incorporated into the design and construction of physical facilities that NSF supports.

7. Intensify the evaluation of current program results and invest in long-term evaluation of program effectiveness in increasing student learning; and, codify and publish what is known about effective practices – to aid in future improvement of student learning, to inform institutional reform efforts, and to provide a validated base for wider and more effective dissemination as well as for future research in learning.

8. In all it does to enhance undergraduate education, NSF must undertake educational improvement as a long-term and continuous process requiring consistent attention and investment. NSF should avoid a "strike force" approach, rushing in to do something and then quickly leaving a program area.

9. Indicate consistently the significance of undergraduate education in presenting the NSF's priorities; use the language of *learning* as contrasted with *teaching*; and lift up *discovery* as a concept uniting education and research.

B. Aggressively improve undergraduate SME&T education through a variety of funding programs, beyond the base recommended in the Neal Report. Virtually a doubling of resources will be needed to erase the gap between present funding of those programs recommended as compared with the totals that report urged, and to extend the benefits of those programs to *all* SME&T students.

1. First priority must be given to allocation of enhanced resources to the activities of the Division of Undergraduate Education (DUE) and to the undergraduate part of the *Alliances for Minority Participation* program in the Directorate for Education and Human Resources (EHR).

2. The Foundation should encourage the research directorates to expand the allocation of their resources to discipline-oriented and interdisciplinary research-related educational activities that integrate education and research and that promote sharing the excitement of, and engagement in, research with undergraduates – with additional emphasis on primarily undergraduate institutions.

In all of its undergraduate programs, NSF should put emphasis on implementation of what is known to work, on genuine institutional change, on propagation of validated good practices, and on sustainability of hard-won improvements. Priorities should be revisited over time and modified as necessary on the basis of continuing assessment and evaluation of program results.

For EHR programs, we recommend expanded support through:

- Implementation of both current and new strategies within the *Undergraduate Faculty Enhancement* program to promote excellence in faculty educational activities and related scholarship; to expand projects to educate graduate students in effective ways to help students learn; and to provide workshops for new faculty at all kinds of institutions, to help them facilitate effective student learning.

- The *Institution-wide Reform* initiative, which can engage every directorate in the Foundation in support of disciplinary and departmental changes that will improve student learning.

- The reach of the *Advanced Technological Education* program into the broad and diverse domain of technology education through its fostering and support of collaborations, alliances, and centers in which two-year colleges, four-year colleges, universities, and industry develop and sustain strong educational synergy that builds on basic educational principles to engender problem-solving and lifelong learning skills, as well as provide quick response to industry's educational needs.

- Programs stimulating and focusing on the scholarship of individual faculty members, like DUE's *Course and Curriculum Development* and *Instrumentation and Laboratory Improvement* programs – whose projects can build a base for improvement of undergraduate SME&T education, generate wider enthusiasm for change, and thus lay the necessary foundation for future institution-wide reform.

- K-12 teacher preparation projects – especially through the *NSF Collaboratives for Excellence in Teacher Preparation* program, where we would recommend funding only projects that *clearly* incorporate the principles of effective SME&T education (in V, VI, and VII above) and show promise of reaching a larger fraction of those entering the profession. And, through

- Expansion of the undergraduate part of the *Alliances for Minority Participation* program.

For the research directorates of the NSF we recommend:

- Expanded support of the *Research in Undergraduate Institutions* and *Collaborative Research at Undergraduate Institutions* programs, which strengthen faculty research capability and the involvement of undergraduate students in predominantly undergraduate institutions. The *Research Opportunity Awards* component of the former is an example of effective address of the needs of faculty in undergraduate institutions by its support of their work in research-rich environments.

- Development of tracks within the *Research Experiences for Undergraduates* program that carry funding not only to where the *research* traditionally is done (i.e., to research-rich institutions, as the program does now) but to where the majority of the *students* are (i.e., to two-year and four-year institutions).

- Adaptation of the *Career Advancement Awards* programs (presently limited to women and underrepresented minorities) to the needs of all faculty members, by supporting a wider variety of shifts in career emphasis (e.g., from disciplinary research to educational research; from instructional research to disciplinary research) than at present.

- Expansion of the *Faculty Early Career Development* program (the "Career" program) from concentration on junior faculty to include support of faculty in mid-career who have strong records of research *and* teaching – to further strengthen the ties between their research efforts and instructional activities, and to illuminate their service as role models for junior faculty and graduate students.

- Modification of the *Grant Opportunities for Academic Liaison with Industry* (GOALI) program, and its well-conceived industry-university partnerships, to include wider faculty participation in programs that will enhance their familiarity with industry's expectations of the background and qualifications of potential SME&T employees.

And, for the whole set of NSF's directorates, we have four cross-cutting recommendations. All of the Foundation's grant-making units should:

- Continue their support of strong activities to correct underrepresentation of women, minorities, and persons with disabilities among students and faculty at the undergraduate level.

- Expand support of the research agenda in human learning at the undergraduate level (see A.5 above).

- Support outreach activities that bring SME&T to the general public through science museums, television, demonstrations in shopping malls, as SME&T education is dependent ultimately on the general public's support for science and technology.

- Consider funding mechanisms that both assign responsibility and provide incentives and rewards for achieving educational excellence for undergraduates not just to individuals, but to whole departments and entire institutions.

C. Provide additional leadership for change in undergraduate SME&T education, beyond program funding, specifically:

1. Together with other major players (such as the NRC, AAAS, ERIC, and the National Library of Medicine), explore the establishment of a national electronic system for validating and disseminating successful educational practices.

2. Support those changes occurring in the faculty reward system that will enhance the quality of undergraduate education, e.g., by convening groups of administrators and faculty to deal with issues such as evaluation of instruction and student learning in SME&T.

3. Provide specific problem-solving training sessions for faculty across institutions, in topics such as how to do inquiry and collaborative learning in large "lecture" classes, how to assess learning outcomes, and how to document learning gains at the departmental and institutional levels.

4. Implement activities that will further engage private industry, business, and foundations in supporting improvement of undergraduate SME&T education.

Our final recommendation is that the National Science Foundation accept leadership of the efforts necessary to implement the thirteen sets of recommendations above.

"And gladly wolde he lerne, and gladly teche." [39]

References and Notes

1. National Science Board, Task Committee on Undergraduate Science and Engineering Education, Homer A. Neal (Chairman), *Undergraduate Science, Mathematics and Engineering Education; Role for the National Science Foundation and Recommendations for Action by Other Sectors to Strengthen Collegiate Education and Pursue Excellence in the Next Generation of U.S. Leadership in Science and Technology,* (Washington DC: National Science Foundation, 1986, NSB 86-100).

2. National Research Council, Center for Science, Mathematics, and Engineering Education, *From Analysis to Action: Undergraduate Education in Science, Mathematics, Engineering, and Technology*, Report of a Convocation (Washington, DC: National Academy Press, 1996), paraphrased from page 4.

3. National Research Council, *From Analysis to Action: Undergraduate Education in Science, Mathematics, Engineering, and Technology*, a National Science Foundation/ National Research Council Convocation, April 9-11, 1995. See also the Challenge Paper for this NRC/NSF Convocation (Washington, DC: National Academy of Sciences, 1995).

4. National Science Foundation, *Contributions of the Social Sciences to the National Science Foundation Review of Undergraduate Education*, A Workshop, February 22, 1996. See Draft Report (March 6, 1996).

5. Wingspread Group on Higher Education, William E. Brock (Chair), *An American Imperative: Higher Expectations for Higher Education*, (Racine, WI: The Johnson Foundation, 1993).

6. National Science Board, *Toward the Next Century: The State of U.S. Science and Engineering* (NSB 94-24) February 1994.

7. Executive Office of the President, Office of Science and Technology Policy, *Science in the National Interest* (August 1994).

8. Elaine Seymour and Nancy Hewitt, *Talking About Leaving: Factors Contributing to High Attrition Rates Among Science, Mathematics, and Engineering Undergraduate Majors* (Boulder, CO: Bureau of Sociological Research, University of Colorado, 1994).

9. David Goodstein, "Scientific Elites and Scientific Illiterates," Sigma Xi, *Proceedings of a Forum on Ethics, Values, and the Promise of Science* (New Haven, CT: Sigma XI, February 25-26, 1993).

10. Three public hearings were convened at the National Science Foundation to invite previously prepared public comment on the issues. The first hearing focused on "disciplinary perspectives" on undergraduate education, and featured testimony from faculty and officers in scientific societies (October 21, 1995). The second hearing provided a forum for "institutional perspectives," and gave us a broad perspective on major changes, actual and needed, from primarily college and university presidents (October 23, 1995). The third hearing was for employers of new graduates, and provided testimony from leaders in business, industry, and a public school superintendent employing new teachers.

11. Luther S. Williams, *Charge to the Subcommittee for Review of Undergraduate Education in Science, Mathematics, Engineering, and Technology*, Advisory Committee to the Directorate for Education and Human Resources, National Science Foundation (June 1995). [Copies available on request; See also Volume II.]

12. The National Research Council announced a "Year of National Dialogue" following the conclusion of the April 1995 Convocation jointly sponsored by NRC and NSF (see reference 3). In order to facilitate this dialogue, the NRC and NSF sponsored four regional symposia underwritten by the Exxon Education Foundation. The first was hosted by the University of Michigan (October), the second by GTE Corporation in Boston (November), the third by the Johnson Space Center in Houston (January), and the fourth by Pomona College in California (February).

13. The NSF sponsored focus groups in support of this review. They were conducted by SRI International under contract with the National Science Foundation. They were held in four locations - Chicago, San Francisco, Atlanta, and Boston during the period October 1995 to April 1996. Nine distinct focus groups were conducted in each region, split between two nearby institutions of higher education. Each focus group was drawn from one of: current students (SME&T majors or non-majors), recent graduates (in SME&T or non-SME&T), students' parents, employers of recent graduates, teacher graduates, and employers of new teachers.

14. Alexander W. Astin, *Undergraduate Science Education: The impact of different college environments on the educational pipeline in the sciences: Final Report* (Los Angeles, CA: Higher Education Research Institute, Graduate School of Education, University of California at Los Angeles, 1992).

15. John I. Goodlad, *Teachers for Our Nation's Schools* (San Francisco, CA: Jossey-Bass Inc., 1990).

16. Tora K. Bikson and S.A. Law, *Global Preparedness and Human Resources, College and Corporate Perspectives*, RAND (Santa Monica, CA: RAND, 1994).

17. Barry Cipra, "Calculus Reform Sparks a Backlash," *Science* (February 16, 1996), page 901.

18. National Science Foundation, Division of Undergraduate Education, Directorate for Education and Human Resources, *Institution-Wide Reform of Undergraduate Education in Science, Mathematics, Engineering, and Technology: Program Announcement and Guidelines* (NSF 95-127).

19. Gene Wright, Andersen Consulting, in "Consulting Giant's Hot Offer: Jobs, Jobs, Jobs," *The Wall Street Journal* (March 16, 1996), page B6.

20. Malcolm Getz and John Siegfried, *Costs and Productivity in American Colleges and Universities*, Part III in *Economic Challenges in Higher Education*, A National Bureau of Economic Research Monograph (Chicago, IL: The University of Chicago Press, 1991) pp. 261-392.

21. U.S. Department of Education, National Center for Education Statistics, *The Condition of Education: A Statistical Report* (Washington, DC: U.S. Dept. of Education, NCES 94-149 and NCES 95-273: 1994 and 1995 editions).

22. Clifford Adelman, *The New College Course Map and Transcript Files: Changes in Course-Taking and Achievement, 1972-1993* (Draft Manuscript, Fall 1995)

23. The Carnegie Foundation for the Advancement of Teaching, *A Classification of Institutions of Higher Education: A Technical Report, 1994 Edition* (Ewing, New Jersey: California / Princeton Fulfillment Services, 1445 Lower Ferry Road).

24. American Association for Higher Education, *Forum on Faculty Roles and Rewards, and Teaching Effectiveness.* AAHE publications addressing the Forum's concerns include: Jon Wergin, *The Collaborative Department: How Five Campuses Are Inching Toward Cultures of Collective Responsibility* (1994); Ernest Lynton, *Making the Case for Professional Service* (1995); and Robert Diamond and Bronwyn Adam (editors), *The Disciplines Speak: Rewarding the Scholarly, Professional, and Creative Work of Faculty* (1995). (Washington, DC: One Dupont Circle). Further information is currently available at the URL: http://www.ido.gmu.edu/AAHE/welcome.html.

25. William A. Wulf, "Warning: Information Technology Will Transform the University," *Issues in Science and Technology*, (Summer 1995), pages 46-52.

26. Eli M. Noam, "Electronics and the Dim Future of the University," *Science*, (October 13, 1995), pages 247-249.

27. The enrollment data were obtained from data tapes prepared by the National Center for Education Statistics (U.S. Dept. of Education), available through the NSF CASPAR data system, using the 1994 Carnegie *Classification of Institutions of Higher Education* (ref 23).

28. The data on the number of students enrolled in undergraduate SME&T courses, and the number of these courses offered, come from special tabulations of data collected in the *1993 National Study of Postsecondary Faculty*, by the National Center for Education Statistics. The tabulations were done by Pelavin Associates of Washington DC.

29. These data are collected by NSF (Division of Science Resources Studies). They were tabulated from raw data for this review. See *Federal Support to Universities, Colleges, and Nonprofit Institutions: Fiscal Year 1992* (Surveys of Science Resources Studies, Detailed Statistical Tables, NSF 94-324). These data are the funds obligated during FY 1992, as reported by 15 federal agencies to the NSF. The totals are for all fields of science and engineering.

30. The data on post high school graduation college activity are drawn from the National Center for Education Statistics Panel Surveys of the high school classes of 1980 and 1982, called *High School and Beyond.*

31. The National Association of Scholars, *The Dissolution of General Education: 1914-1993* (Princeton, NJ: March 1996). An executive summary is available on the Internet at the NAS home page.

32. Letter dated June 14, 1995 from Dr. Luther Williams to a widespread mix of educators, administrators, and employers in the science, mathematics, engineering, technology, and higher education communities. See Volume II for the full text of this letter.

33. National Science Foundation, Division of Undergraduate Education, Directorate for Education and Human Resources, *Project Impact: Disseminating Innovation in Undergraduate Education, Conference Proceedings* (May 1994, NSF 95-69).

34. Recommended Standards for high school graduates are found in the following carefully crafted reports: (1) National Research Council, National Committee on Science Education Standards and Assessments, *National Science Education Standards* (Washington, DC: National Academy Press, 1995); (2) American Association for the Advancement of Science, Project 2061, *Benchmarks for Science Literacy* (New York, NY: Oxford University Press, 1993); (3) National Council of Teachers of Mathematics, *Curriculum and Evaluation Standards for School Mathematics* (Reston, VA; 1989).

35. Sheila Tobias, *Revitalizing Undergraduate Science: Why Some Things Work and Most Don't* (Tucson, AZ: Research Corporation, 1992).

36. Robert Rosenthal, "Teacher Expectancy Effects: A Brief Update 25 Years after the Pygmalion Experiment, *Journal of Research in Education* (Spring 1991), pp. 3-12; results of a meta-analysis of 448 studies. See also, Robert Rosenthal, "*Pygmalion in the Classroom;* Teacher Expectation and Pupils' Intellectual Development)New York NY: Holt Rinehart & Winston, 1968).

37. Lewis Thomas, *Late Night Thoughts on Listening to Mahler's Ninth Symphony*, (New York, NY: Viking Press; 1993), page 155.

38. Chancellor Karl Pister, "Renewing the Research University," *University of California at Santa Cruz Review* (Winter 1996).

39. Geoffrey Chaucer, *The Canterbury Tales; Prologue*. Line 308.